Survival
of the
Fattest
The Key to Human Brain Evolution

Survival
of the
Fattest
The Key to Human Brain Evolution

Stephen C. Cunnane
Canada Research Chair in Brain Metabolism
Research Center on Aging
Université de Sherbrooke
Québec, Canada

World Scientific

NEW JERSEY · LONDON · SINGAPORE · BEIJING · SHANGHAI · HONG KONG · TAIPEI · CHENNAI

Published by

World Scientific Publishing Co. Pte. Ltd.

5 Toh Tuck Link, Singapore 596224

USA office: 27 Warren Street, Suite 401-402, Hackensack, NJ 07601

UK office: 57 Shelton Street, Covent Garden, London WC2H 9HE

British Library Cataloguing-in-Publication Data
A catalogue record for this book is available from the British Library.

First published 2005
Reprinted 2006

SURVIVAL OF THE FATTEST
The Key to Human Brain Evolution

ISBN 981-256-191-9
ISBN 981-256-318-0 (pbk)

Printed in Singapore by B & JO Enterprise

To Clare, Sara and Monique, who make it all worthwhile.

Contents

Survival of the Fattest

Foreword

In this book, Stephen Cunnane has turned the search for human origins from loose speculations into science. The accepted view of humans evolving a large brain on the savannahs of Africa was sparked by Raymond Dart's words in 1925 when he said that 'open country with occasional wooded belts and a relative scarcity of water, together with a fierce and bitter mammalian competition, furnished a laboratory such as was essential to this penultimate phase of human evolution'. Since then paleaoanthropologists have stuck to what became the "savannah hypothesis" with almost religious fervour. This hypothesis lacked any evidence base. It never explained why every single land-based mammal proportionately lost brain size in relation to body size as they evolved larger bodies. The loss was universal starting from brain representing greater than 2% of body weight for small mammals and primates, shrinking to less than 0.1% of body weight for larger mammals. The 350 gram brain of a one ton rhinoceros pails into insignificance compared with the 1,800 gram brain of the dolphin or the 8 kilogram brain of the sperm whale. This universal and massive loss of brain size in herbivorous land species, occurred despite what must have been successful competition for food against the large carnivores and other mammals. Otherwise, the herbivores could not have evolved larger and larger bodies.

What Stephen has done is to come up with strong support for an alternative, evidence-based view that humans evolved in a coastal ecosystem. A coastal origin was first articulated by Sir Alistair Hardy in the 1960s. Although it found several supporters, it has been roundly rejected by palaeoanthropologists. Yet, there is not a single piece of evidence or even a testable hypothesis to explain how *H. sapiens* escaped the trap of diminishing relative brain capacity with increasing body size on land. Stephen provides evidence. He injects two essential criteria for human cerebral evolution: the requirements for (i) more energy (ii) and more refined structure. He then presents the evidence as to how these two criteria would have been met. The energy cost for human brain growth, which reaches 70% of all the energy the mother ploughs into her fetus during the brain growth spurt, is guaranteed by fat stores unique to humans amongst the primates. Comparative studies first demonstrated the universality of docosahexaenoic acid (DHA) in neural signalling systems of the fish, amphibia, reptiles, birds, mammals and primates alike. Science then showed that deficiency of brain DHA leads to degenerative processes, reduced cognitive and visual function. Conversely, added DHA, even in well nourished Norwegian women, leads to increases in IQ in children. Genomic research has explained that DHA is responsible for transcribing over 170 genes required in brain development. That is, the structural and bio-motive requirement is met by the same fatty nutrients that formed the first neurones in the sea some 500 million years ago.

A second critical feature of the book is that it turns attention away from the macho image of killing or scavenging for food, or even the most recent idea of endurance running to compete for food, towards the singular and paramount importance of the mother. It is an uncomplicated fact that the nourishment of the mother would have determined the forward or backward evolution of the brain. This brings me to the superiority of Stephen's thesis over the savannah proposals. Nowhere in the latter is there a notion of the significance of "brain-selective nutrients". Food is food and that is it. The universal collapse of relative brain size in land-based species despite the advance in massive body size clearly demonstrates there are different requirements for body growth on the one hand (protein and minerals) or for the brain on the other (lipids and trace elements). That fact is so simple, it is strange that it has been ignored. The idea of hominids evolving a large brain by competition on the savannahs contains no predictive power. Yet prediction and test is the hallmark of science and progress. The persuasive evidence presented by

Stephen has the power to predict both forward and backward evolution. His wide view of many facets of evolving humans adds to a compelling story, which has a message not just for our past, but for our present and future. It was a privilege to read this book before its publication.

Professor Michael A. Crawford, PhD, CBiol, FIBiol, FRCPath
Director, Institute of Brain Chemistry and Human Nutrition
North Campus, London Metropolitan University
166-229 Holloway Road
London N7 8DB
Tel 02079734869, Fax 02077533164
www.londonmet.ac.uk/ibchn

Preface

How Did the Human Brain Evolve?

Four elements combine to influence evolution of any biological feature – time, pre-existing rudimentary capability, genetic variation amongst individuals, and some change in the environment. The first two of these elements are not contentious in relation to human brain evolution. Humans today are the living proof that sufficient time - at least five million years - was available to evolve brains that are three fold larger compared to either other living primates or our earliest forebears.

The great apes are the closest genetic relatives to humans and clearly have some rudimentary capability to learn and solve problems. Higher cognitive function was therefore more likely to occur in an ape, which already had a relatively large and advanced brain, than in a cow or even a tiger, both of which have relatively small brains compared to primates. Humans and chimpanzees last had a common ancestor 6 to 8 million years ago and it seems likely that the earliest human forebears had cognitive capabilities broadly similar to that of present day chimpanzees. Hence, two of the necessary elements – sufficient time and preexisting rudimentary cognitive capability in the last common ancestor of humans and apes - are beyond reasonable dispute.

However, there is little agreement on the relative role of the other two elements that participated in human brain evolution: genetic variation and change in the environment. In essence, this is a nature versus nurture debate, the origin of which is at least a century old.

This debate is now hotter than ever as molecular biologists artificially manipulate genes and brain function in ways unimagined even a decade ago.

Despite clear differences in body morphology and brain capacity, there are surprisingly few differences in the genomes of humans and chimpanzees. At a minimum, differences in genes controlling aspects of brain development including neurogenesis, neurotransmitter receptor expression, and myelination are likely to have participated in the emergence of advanced cognitive function in humans. The ways in which these processes operate and are controlled by genes are being intensively explored but have not yet revealed how the cognitive differences between humans and apes evolved.

Humans have the potential for intelligence but, in acquiring that unique specialization, appear to have paid a price in terms of increased vulnerability of the brain, especially during infant development. It takes at least five years before human infants are minimally independent of their parents and even then most cannot survive unaided. In the case of human brain evolution, not only how the additional operating costs of the larger brain were met needs to be accounted for but also how, simultaneously, human evolution tolerated increasing brain vulnerability during infant development.

Neurodevelopmental vulnerability is the clearest indication that an interplay between genes and the environment contributes to the exquisite functional potential of the human brain. That interplay constitutes Darwin's *'conditions of life'* and made possible human brain evolution, just as it makes possible evolution of any other attribute. But how do genes interact with the environment to permit evolution of cognitive skills that have no apparent survival value while simultaneously creating a long, vulnerable period of early development?

Brains in general require a lot of energy so a bigger brain is even more expensive than usual. A large, metabolically expensive, and developmentally vulnerable brain is impressive enough, but *hominids* destined to become humans didn't just evolve larger brains; they also evolved fat babies[1]. It may seem only moderately difficult to meet the energy and nutrient requirements for a larger brain, but added to the human brain's remarkable early development is the simultaneous accumulation of considerable body fat before birth. The fat accumulating on a healthy human fetus as it approaches birth is not present in other primates but is, in fact, a prerequisite for full development of advanced brain function in human adults.

***

Since the time of Raymond Dart, the environmental catalyst for early hominid divergence from the last common primate stock has been thought to be a hot, dry climate that created the savannahs in East Africa 4 to 5 million years ago. Those conditions were thought to have forced a four-legged, climbing ape to become earth-bound and search for food while becoming *bipedal* (walking on two legs). Now, after well over fifty years of muted grumbling, even from former supporters, the *Savannah Theory* is transforming towards the *Woodland Theory*, in which the climate of East Africa is now viewed to have been less extreme and forests were more abundant. Hominids are still seen as evolving from an arboreal ancestor but they stayed in the woodlands, more-or-less like today's other great apes.

Some of the questions I want to address in this book are: If both the human genome and the environment hominids inhabited were so similar to those of the great apes, especially the chimpanzees, how did some hominids go on to acquire such a unique brain? What was the catalyst not only for evolution of a larger and more advanced brain but neonatal body fat? Insufficient dietary energy or nutrients severely challenge both

---

[1]*Hominids* are defined as the branch of primates that became bipedal but did not necessarily form part of the final lineage to humans. *Hominins* are the branch of hominids that led to humans – *Homo (H.) sapiens,* or *Homo sapiens sapiens*. Exactly which of several hominid and then hominin species were the forebears of humans is still unclear. In the interests of simplicity, I will use 'hominid' as a general term for bipedal primates, and will specify the species where known or appropriate. I will also use 'pre-human hominid' to denote the pre-human lineage.

brain development and fat deposition in the human fetus today, so did these same deficits curtail evolution of both advanced brain function and fat deposition in the fetuses and infants of non-human primates? This book is about the 'conditions of life' that, starting from an already successful primate blueprint, permitted only pre-human hominids to avoid or squeeze through this double-pronged evolutionary bottleneck.

Acknowledgements

A number of people helped me with this book. Chronologically, they start with Leo Standing at Bishop's University, who triggered my interest 30 years ago in one of his undergraduate courses. He endured several oral and written versions as my ideas about human brain evolution developed. My PhD mentor, David Horrobin, encouraged me by publishing my first paper on this subject in *Medical Hypotheses,* the journal he edited. Many discussions about human brain function with Michael Crawford led me to thinking that evolution of body fat in infants was the unique human solution that broke the barrier to brain expansion experienced by other primates. Several meetings with Elaine Morgan over the years reinforced my feeling that, more than other distinctive aspects of human morphology and physiology, brain evolution was likely to have been the main beneficiary of shore-based evolution in humans. Her editorial skills are also much appreciated. During more than a decade of overlapping research interests concerned with polyunsaturates and brain development, the critical intellect and warm friendship of Tom Brenna has been very gratifying. Kathy Stewart has been very supportive with many discussions and forthright comments on the manuscript. I am thankful to the members of my research team and to colleagues and acquaintances the world over who have encouraged the evolution of the Survival of the Fattest. Amongst them, Mary Ann Ryan stands out for her patience and skill and without whom our research would be a shambles. Notwithstanding their efforts and support, I am responsible for the shortcomings and errors that remain.

Part 1

The Human Brain: Unique Yet Vulnerable

Chapter 1

Human Evolution:
A Brief Overview

What is Evolution?

Evolution involves gradual change in the form and function of organisms. Since one's phenotype is coded for in DNA, either the genotype itself or the way in which the genotype codes for the phenotype must change in organisms that evolve. In organisms such as bacteria in which the reproductive cycle is short, changes in DNA can bring rapid changes in phenotype; in organisms with a long development time, these changes take longer. Though it is an extremely large molecule and has to replicate many times as the body's billions of cells divide, DNA is a highly reliable code. Still, changes in the nucleotide sequence of DNA do occur. If the change in DNA is too extreme or affects a critical site in the genome, the egg may not get fertilized or the embryo may not be viable. If it is too slight, the change may not show up in the phenotype of the offspring.

Each species is said to have a single set of genes or *genome*. But the genome of one individual is slightly different from everyone elses. No single individual has the ideal model genome for that species. Defective copies of a gene in one parent may be silent because the other parent's copy for that gene is correct. Even matching but defective sections of DNA in both parents may not matter because those sections do not code for genes. The basis of evolution is that there is just enough non-lethal change in just the right part of DNA.

Climate (temperature, day length, humidity, radiation), predators, food supply, water supply and population density are all factors that constitute the environment. Part 1 of this book outlines the influence of food components such as brain selective minerals and docosahexaenoic acid on the development of a normal human 'neurological' phenotype; i.e. the physical, biochemical and physiological constituents of normal brain and behavioural development within an individual. Culture also influences brain development but, again, mostly by affecting gene expression within the individual.

The issue related to human brain evolution is – how do dietary, cultural or hormonal influences in an adult affect the genotype or gene expression in that adult's offspring? Nutrient deficiency affecting brain function in a parent doesn't impair brain function in the child unless the deficiency changes gene structure or unless the child also consumes the same diet. Parents' habits or cultural milieu have an important influence on habits of their children but how does that change the child's genotype? If food or the environment don't change DNA in the second generation, evolution doesn't get started.

Random mutations in DNA are exempt because, by definition, they will happen randomly; their timing, frequency and point of impact on DNA are not predictable. Random mutations are also extremely unlikely to hit specifically the right genes and in the right way that a phenotype evolves rather then dies. However, different environments could influence the occurrence of random mutations.

Environments can also change dramatically but they don't change randomly; here today, gone tomorrow. A rare but dramatic environmental change such as a volcano spewing lava, ash and darkness for months, occurs too suddenly for most organisms affected by the sudden change to adapt. There are of course exceptions but, generally speaking, environments change gradually and it is the role of these gradual changes that are of interest here.

Natural Selection : Need Versus Blind Consequence

Charles Darwin coined the influential term used to describe how evolution occurs – *natural selection*. Natural selection involves four essential ingredients: (1) time, (2) inherent variability between individuals (3) environmental change, and (4) some preexisting rudimentary capacity or adaptation towards the characteristic in question. These four variables allow characteristics (phenotypes) to evolve. With sufficient time and preexisting capacity, environmental change tips inherent variability in a certain feature in a particular direction, e.g. rudimentary standing ability gets tipped towards committed instead of opportunistic bipedalism. Without differences in these *'conditions of life'*, there would be less variability within species and less for natural selection to work with.

Darwin recognized that inherent variability in a characteristic within a species was necessary but was not sufficient to stimulate evolution. The crocodile has variability in many characteristics but has not evolved substantially because no stimulus - no conditions of life - have provoked change in favour of those with larger or smaller jaws, etc. The environment creates the conditions of life, which may be constant or may suddenly vary. As Richard Lewontin (Harvard University) notes, conditions of life vary with the organism. A backyard pond may look like just a pond to a human but the conditions of life for insects in that pond change dramatically if it becomes inhabited by a frog or if we drop in a couple of gold fish.

Climate change is an important condition of life that creates what is commonly called *selection pressure*. Selection pressure implies that evolution can be pushed, whether by environmental change or by a certain need. But need (real or imagined) cannot condense the time required to select and mate just the right individuals to withstand the imposed change of environment. All species follow the same rules of natural selection. Those individuals who, by chance, are most suited to withstand the climate change get to take advantage of their better suited genotype. Climate change can't force a change in genotype, no matter how useful the survival value of altering a particular feature.

Selection pressure creates a common 'cart before the horse' problem in most explanations of the origin of the human brain because an enlarged brain is widely thought of as having a significant survival advantage: if you have a bigger brain and are smarter, then you can make better tools, hunt better, and find a similarly better-adapted mate. That offers you and your offspring a better chance to survive but it carries the disadvantage that you now need to find more food in order to keep your enlarged brain functional. Unless more food remains available, your enlarged brain cannot be sustained, and you end up with no more advantage than those with smaller brains.

It is difficult to resist invoking the need for a better quality diet or the need for weapons and organized hunting to support the nutritional requirements of a large brain. Humans do need a better quality diet now that we have a large brain. A change in the conditions of life, i.e. a better quality diet, was also necessary to get a larger brain. However, everything revolves around the point that hominids couldn't need the complex and sophisticated behaviours permitted by a large brain (hunting, language, tool making, etc.) in order to get a large brain. This is the catch-22 – how did certain hominids meet the requirements of brain expansion without needing the skills that arose with brain expansion?

The Struggle for Existence

Over the course of at least five million years, humans have evolved out of the mold of an ape-like ancestor. Along the way, short hairy bodies, crude stone tools, medium-sized brains, and simple skills gave way to taller, relatively hairless bodies, body fat, unusually large brains, and many languages and cultures. Seemingly limitless creativity and technological potential became interwoven with a combination of carelessness and aggression that at times severely strains the common good. Early human ancestors have long been thought of as maladapted intermediates that willed their way to a better life. Humans are considered to be the beneficiary of that effort and drive and, as such, are still widely viewed as the pinnacle of primate evolution.

Sir Grafton Elliot Smith, a leading British anthropologist in the 1930s, promoted the idea that humans evolved by overcoming adversity and sloth: *'[Our ancestors] were impelled to issue forth from their forests and seek new sources of food and new surroundings on hill and plain where they could find the sustenance they needed ... The other group, perhaps because they happened to be more favorably situated or attuned to their surroundings, living in a land of plenty, which encouraged indolence in habit and stagnation of effort and growth, were free from this glorious unrest, and remained apes, continuing to lead very much the same kind of life* [as the ape-human ancestor]. *While man was evolved amidst strife with adverse conditions, the ancestor of the Gorilla and Chimpanzee gave up the struggle for mental supremacy because they were satisfied with their circumstances'*.

With *'the struggle for existence'*, Charles Darwin planted the seed nearly 150 years ago that evolution is rooted in adversity and strife and leads to a place or physical condition where there is less of a struggle; evolution is a process of perfection. Grafton Smith reinforced this view of human evolution with the *'struggle for mental supremacy'* and *'evolution amidst strife with adverse conditions'*. Ever since, explanations for human evolution have consciously or unconsciously integrated a process of success against adversity: the adversity of a harsh climate, competing against predators, finding sparse food and becoming bipedal and large-brained.

The concepts of struggle and survival seem to be imposed on the process of evolution; evolution has been humanized in the context of the Victorian era work ethic. Certainly an individual or a small group may struggle to survive but a species doesn't struggle to evolve. Human brain evolution didn't occur under adversity or strife any more so a million years ago than it does today. Hundreds of millions of people today live with constant adversity and struggle to feed themselves and their families. Does that mean that as a species, *H. sapiens,* is undergoing a struggle for existence?

Evolution occurs because of opportunity, not necessity. An organism's existence is challenged or it isn't. If it is challenged, and if it and enough of its progeny adapt, the species survives; if not, it becomes extinct. Evolution occurs when there is sufficient variation in a biological feature that, when provoked by an environmental stimulus of sufficient magnitude, favours reproduction of those with one variant of that feature while disfavouring the others. The feature could be tooth shape, form of locomotion, brain size, or anything else. In principle, every aspect of our form and function varies from individual to individual and is susceptible to natural selection.

The heart is recognizably similar across all mammals. According to natural selection, this indicates that despite many different environments and body plans, the four-chambered mammalian heart works well. It has undergone little or no change because it perceives no real difference in its own environment, no matter whether it is hot or cold, wet or dry, or whether wings, fins or legs are used for locomotion.

How mammals move around is a different matter. Despite similar musculo-skeletal organization, mammals possess a variety of locomotory appendages including winged or webbed forelegs, two or four functional legs, one to five toes, fins, etc. Thus, the musculo-skeletal system is highly malleable by the environment.

What about brains? Grafton Smith's *indolent, stagnant great apes* have somewhat disproportionately large brains but not nearly so much as humans. This means that unusual circumstances were required to make brains expand in pre-human hominids. Importantly, there was enough natural variation in primate brain size that, with the right stimulus, it could begin to disproportionately increase in certain hominids.

What I wish to challenge is the deeply ingrained idea that bigger brains were better for survival and that in order to survive, hominids had to do certain things that needed a bigger brain, such as find meat, develop tools and language, and become social animals with culture. Without possessing particularly large brains, the great apes have survived much longer than humans. Some hominids survived for upwards of two million years so, despite smaller brains, in evolutionary terms, they were still successful. They didn't have to become human or acquire larger brains to be evolutionarily successful.

Natural selection has no goals. Each change in form or function either benefits or impedes the individual. What the individual's anticipated descendents could potentially do if a particular feature were refined, improved or abandoned is of no concern. Indeed, natural selection wouldn't know which way to plan evolution because it cannot anticipate when or where DNA mutations might strike, or when volcanoes might explode and alter the environment of large areas for a prolonged period. Hence, the fact that early pre-human hominids had, at best, modestly improved cognition, were crude tool users, and probably had no more than simple language, was not an impediment to their successful reproduction and propagation.

H. erectus appears to have lasted for over a million years without substantial change in physique. Humans *per se* have lasted about 100,000 years but, despite unsurpassed intelligence, are constantly at risk of destroying themselves by a variety of means including warfare, pollution, and carelessness. Neanderthals had larger brains than *H. sapiens* but although they existed longer than have humans, ultimately, they did not survive as a species. Hence, humans should not be viewed as the pinnacle but may indeed be the ignominious *Homo*.

Brain evolution can be looked at in a different way than it has so far; brain expansion was neither necessary for nor advantageous to survival or evolution of hominids. University of Michigan paleoanthropologist, Milton Wolpoff, put it succinctly: '*you could never make the case that intelligence is the best strategy. Over the past 20 million years, monkeys have been much more successful than apes, even though apes have bigger brains and are more intelligent by any measure*'.

Like Grafton-Smith, James Shreeve notes that gorillas have a simple existence; they do little except eat, sleep and play, yet they are amongst the most intelligent of animals. Shreeve and Grafton Smith therefore contrast starkly in their appreciation of the relation between a life of leisure and the origins of intelligence. The potential for intelligent behaviour has clearly and uniquely arisen in humans but it also poses many challenges and has a dark side. It is important to remain cognizant of both the benefits and the pitfalls of higher intelligence if its origins are to be fully understood.

Fossil Evidence

Interpretation of how humans evolved is based on two related types of physical evidence. First, there's what has been learned about the similarities and differences in form and function of humans and primates most closely resembling humans. Since at least Darwin's speculation about human origins in Africa, it has been a reasonable assumption that the similarities between chimpanzees and humans indicate a common ancestor. Human genes are now known to be at least 99% identical with those of chimpanzees. This means that the genetic blueprint for humans and chimpanzees is almost identical but it doesn't mean that all these genes are expressed or operational to the same degree in both species.

The other key window on human origins is the nature, age and circumstance of hominid fossil beds and their contents. Interpreting fossils requires detailed knowledge of the shape and size of the bones and teeth not only of reasonably closely related species like humans and chimpanzees but of many other species as well. Only very rarely are fossilized bones of the skeleton or skull found intact. Teeth are better preserved because of their extreme hardness. Nevertheless, there are many intermediate forms of jaw shape and skull size that could account for a specific tooth shape. Large flat teeth and heavy jaws suggest repeated chewing of plant material while large, pointed canines suggest meat consumption. Dating the age of fossils has become more sophisticated and reliable, which has also greatly helped locate fossils in time if not always in their relationship to living humans or to each other.

Still, by far the biggest challenge is that there are insufficient fossils with which to unambiguously reconstruct the pre-human form. It is rare to find fossils that are intact and in which the bone shape is clear and uncontroversial. Most are found as fragments so it is a painstaking process to find and reassemble them into their original shape. Rarely can a fairly complete puzzle be reconstructed. This is a particular challenge for those interested in brain evolution because the skull is relatively thin and hence prone to being broken and crushed. The skull also covers a large three-dimensional surface, thereby multiplying the risk that each small error in aligning an incomplete set of fragments during reconstruction will affect the final size and shape of the reconstructed skull.

The two key features of human brain evolution that are available from fossils are the size and the shape of the skull. Finding more than one specimen of the same species and age helps establish how big the brain really was. In the rare and fortunate cases in which relatively complete skeletons and skulls are available, it is possible to compare skull size and shape with the probable posture of that individual.

Humans walk vertically and have large skulls, small jaws and relatively small molars for crushing food. Bigger hominid fossil skulls tend to date from the more recent past and tend to be associated with individuals that walked more vertically and had smaller jaws and teeth. They also tend to be found at sites where there is more evidence of tool making and use of fire.

The earliest hominid fossils generally indicate that hominids became bipedal before they started using tools or before the brain had started to increase very much in size. If this sequence is true, one important deduction that follows (walking preceding tool use preceding a significant increase in brain size) is that evolution of bipedalism could not have depended on larger brain size to coordinate movement and balance.

In addition to the fossil evidence that bipedalism predated significant brain expansion, this sequence makes intuitive sense because tool making requires the hands. If the hands were no longer needed for walking because their owner was bipedal, they would have been available to do other things. Though the tendency is to try to relate as many features to each other as possible, there may not, in fact, be any direct relationship between walking on two feet and brain size or, for that matter, tool use.

Bipedalism may have been advantageous for the hands and possibly for the brain. What makes sense using that premise should be explored but, at the same time, we should not ignore the possibility that they are unrelated. As Claude Bernard, the famous French physiologist and father of experimental medicine, said - we must *'believe and disbelieve at the same time'*: For evolution to occur, the changing feature (bipedalism) had to be tolerated by the whole organism. Realigning the torso from a largely horizontal to a vertical line through the head, back and pelvis is a major change in design. Why this was advantageous (or, in fact, acceptable) to the rest of the body?

Australopithecines

The evolutionary pathway to humans starts in East Africa as the Miocene turned into the Pliocene epoch about 5 million years ago. Five million years earlier in the middle of the Miocene, continental uplift and faulting produced the Rift Valley in East Africa. Earlier in the Miocene, some drier areas already existed in the northern and southern extremes of Africa but towards the end of the Miocene and the beginning of the Pliocene, repeated glaciation in the Northern Hemisphere led to widespread drying of warmer climates, reduced sea levels and fairly widespread replacement of tropical forests by savannah in equatorial Africa.

The Miocene was a period of abundant speciation involving the emergence of new mammals, including the common ancestor of apes and the early variants of hominids. Hominids were the first branch of primates to become irreversibly bipedal, which is considered to be the first and most fundamental anatomical commitment towards becoming human. The eastern side of the Rift Valley was drier and experienced repeated volcanic activity. This is also where many early hominid fossils are found. In contrast, the western side of the Rift Valley remained cooler and wetter and is where the non-human primates prospered.

The best-known Australopithecine skeleton is that of Lucy, who is classified as *Australopithecus (A.) afarensis*, for the Afar region of Ethiopia where she was uncovered by a team led by Donald Johansen. Lucy and others like her were essentially three-foot tall, bipedal apes that lived 3 to 4 million years ago (Figure 1.1). The wear patterns on their teeth suggest a mixed diet of plants, fruit and perhaps a little meat. Paleopanthropologists like Richard Leakey feel that *A. afarensis* was not in the hominid lineage so this controversy leaves the shape of the early pre-human lineage very uncertain. Debate also occurs about whether another 4 million year old Australopithecine, *A. garhi*, was in the line that led to humans. Nevertheless, there is general agreement that Australopithecines left little or no evidence of having been tool users.

One branch of Australopithecines existed 1.3 to 2.7 million years ago and is generally agreed not to have been pre-human (Figure 1.1). These hominids had heavier physiques and larger teeth and are sometimes categorized as a separate genus - *Paranthropus (P. ethiopicus, P. robustus,* and *P. boisei). Paranthropus* species are also known as the 'robust' Australopithecines because they were physically heavier the 'gracile' or slim Australopithecines such as *A. africanus* that existed simultaneously during the period 2.2 to 2.7 million years ago. The heavy jaw structure and large flat molars of *Paranthropus* suggest they were specialized plant eaters consuming a diet that required considerable mastication. *A. africanus* is usually viewed as the most recent Australopithecine leading to *Homo*.

Homo habilis

By the start of the second to last major polar glaciation about 2.6 million years ago, the various Australopithecine species had disappeared, and the markedly larger males and smaller females that characterized Australopithecines disappeared with them. The final major glaciation occurred about two million years ago, marking the beginning of the Pleistocene epoch and the emergence of the first *Homo* species (Figure 1.1).

Homo (H.) habilis (handy man) existed 2.4 to 1.8 million years ago and left the first clear evidence of having intentionally made stone tools. Tools made by *H. habilis* are the simplest and are called Oldowan technology for the Olduvai Gorge in Kenya where they have been found in abundance. Some paleoanthropologists feel that the skull of *H. habilis* shows the first evidence of asymmetry in the brain's left hemisphere that is associated with development of areas for speech and language. Stone tools made by *H. habilis* appear to have been predominantly made by right-handed individuals. These two points stress the emerging links between increasing brain size, manual dexterity and tool making.

It is clear from the fossil record that a little less than two million years ago, *H. habilis* co-existed in some areas for at least 100,000 years with two or three other hominids including *H. rudolfensis, H. ergaster* and possibly early *H. erectus*. Thus, early hominids were not purely a sequence of species heading towards humans but branched out into parallel lines evolving simultaneously. One or two *Paranthropus* species also existed at this time. Different species are generally thought not to share identical ecological niches. Since the stone tool technology developed by *H. habilis* was apparently not shared or acquired by other hominids at that time, it is reasonable to conclude that, though they coexisted temporally, some *Homo* species probably were specialized to distinct diets, habitats and environments.

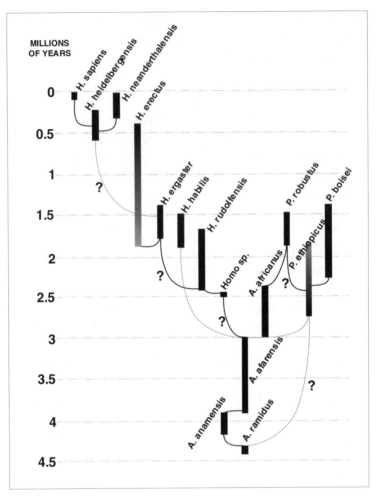

Figure 1.1. The phylogenetic tree of human origins starting about 4.5 million years ago (modified from Broadhurst *et al* 1998). A – *Australopithecus*; P – *Paranthropus*; H – *Homo*.

Homo ergaster

H. ergaster was an African hominid with similarities to *H. erectus*. The best known example of *H. ergaster* is 'Turkana boy', a fairly complete, adolescent male specimen found near Lake Turkana. With the arrival of *H. ergaster*, hominids had gained about 50% in height and were approaching *H. sapiens* in body form. This substantial change in height and the less distinct wear patterns on the teeth suggest that *H. ergaster* ate less plant material and probably had a higher quality diet than *H. habilis*. The usual inference is that more meat was being eaten but the location of Turkana boy's fossils near Lake Turkana means that other food was certainly available, including shellfish and fish.

H. ergaster is credited with two other significant steps towards becoming human. First, they used a new, higher quality stone tool – the so-called Acheulian hand axe. Second, they appear to have developed the first controlled use of fire. Despite the development of more refined stone tools, the cruder Oldowan tools remained in wide use at the same time. Hence, there was no sudden abandonment of the simpler tools in favour of the more sophisticated technology.

This story repeats itself with the monotonous use of highly refined stone tools by Neanderthals despite clear evidence that early humans (the Cro-Magnon) were simultaneously using more advanced tools made from materials other than stone. If the new tools developed by *H. ergaster* were necessary for killing or dissecting animals to be eaten, or for preparing clothing or shelter, one would imagine that such new technology would have been seen as advantageous and would have been more widely adopted.

Stones whose edges were purposely flaked off to leave a sharp edge are almost always viewed as having been tools to cut meat, hides or other materials. However, in some instances these flaked stone tools are more sophisticated than necessary for the apparent task at hand. Interpreting why certain tools were made the way they were, i.e. whether they were used purely for subsistence purposes or whether they possibly had artistic or decorative uses, may therefore need reevaluation.

How these tools were used and what difference they made to the lives of their owners remains enigmatic. Some experts feel that, despite the increase in brain size that seems to be linked to the emergence of the Acheulian tools, nothing in the archeological record shows a real change in lifestyle of *H. ergaster* over *H. habilis*. Hence, increased brain size may have occurred independently of technological change: increased computing power did not necessarily have rapid repercussions on the quality of life of hominids.

Once again, the critical factor is the quality of the archeological record - it may never be clear what skills or material goods were accumulated because few of them were made of substances that fossilize or are otherwise preserved. If it is correct that evolution of a bigger brain was dissociated from increased capacity to make more sophisticated tools, then this is important in interpreting different options for the lifestyles and evolution of early hominids. It also takes some of the focus off needing a larger brain for survival.

Homo erectus

The glaciation ending about two million years ago was followed by the emergence of the globe-trotting *H. erectus*. Uplifting, faulting and extensive volcanic activity characterized this period, leading to confusingly mixed layers of igneous and sedimentary layers with volcanic ash in between. During the Pleistocene, much of the climate of East Africa became hot and dry but there were still massive lakes in equatorial East Africa, including paleo-Lakes Turkana and Olduvai, as well as what was to become present-day Lake Victoria.

H. erectus overlaps entirely in the archeological record with *H. ergaster* and emerged as *H. habilis* was nearing the end of its run as a species. However, *H. erectus* lasted much longer than the earlier hominids and was probably around when the earliest anatomically modern (though behaviourally primitive) humans appeared 300,000 years ago. *H. erectus* seems to have been the first truly adventuresome hominid because their fossils have been found in a wide ranging area extending from Africa to China to Java.

As recently as 25 years ago, many similar hominid fossils were vying for the status of being distinct species. The picture is gradually being consolidated as the variations between these specimens are considered to be insufficient to merit the naming of an individual species or genus. Hence, *Pithecanthropus* or Java Man is now *H. erectus, H. transvaalensis* is now an Australopithecine, and *Singanthropus pekinensis* is now *H. erectus*.

At the moment, at least three *Homo* species preceding *H. sapiens* are now widely recognized (*H. habilis, H. erectus,* and *H. neanderthalensis*).

Besides being bipedal, all *Homo* have other key features in common making them distinct from other primates, including large brains, truly vertical stance, modestly sized jaws located directly under the face, and opposable thumbs. Some experts feel the lumping of hominid specimens under a total of three or four species names has gone too far and that distinctions are now being overlooked for the convenience of an overly simplified lineage.

Homo heidelbergensis

It is unclear whether *H. erectus* led to *H. sapiens* or was a separate terminal hominid branch. Even if *H. erectus* was a separate branch, there are still several distinct hominids between *H. ergaster* and *H. sapiens* (Figure 1.1). These include the 800,000 year old *H. antecessor* recently discovered in Spain, as well as the somewhat younger *H. heidelbergensis*. *H. heidelbergensis* had a moderately large brain and left scattered archeological evidence of social structure and possible controlled use of fire. Still, the types of tools they used were not noticeably improved compared to those in use at least 500,000 years earlier in Africa. This repeats the increasingly familiar disjunction between anatomical evidence for approaching 'humanness', i.e. larger brain size and erect stature, yet paucity of evidence in the archeological record of artifacts indicating greater cognitive capacity.

If such tools were not being made, what circumstances supported evolution of a considerably larger brain? Put another way, if *H. heidelbergensis* did not depend on more sophisticated tools to hunt or survive, what were they eating to sustain the increasing energy needs of the enlarging brain and the gains in body height and weight over *H. erectus* and, especially, over *H. habilis*? Did they not need make more refined tools or were they still not able to? If they didn't need to make sophisticated stone tools, is the notion that they needed their bigger brains to invent tools or weapons for survival still tenable?

H. heidelbergensis shows little evidence for improved tool design but they do show the first apparent anatomical evidence for speech. If this modification in the skull shape of *H. heidelbergensis* means they had speech, then they seem to have been the first hominid in whom language and, hence, verbal communication, was emerging. This would represent the fundamental roots of what is commonly referred to as culture, which is one of the cardinal distinctions between humans and other species. Hence, there is intense interest to date the emergence of speech, language and other early signs of culture.

Homo neanderthalensis

H. neanderthalensis (Neanderthals or Neandertals) is so-named because the original specimens were found in a cave in the Neander River valley of Germany. Neanderthals were the only para-human hominid because although early humans had evolved during the Neanderthal period, DNA evidence tends to suggest that they were distinct and did not interbreed. They lived for about 200,000 years but became extinct about 30,000 years ago.

Based on the size and thickness of the fossilized bones, Neanderthals were bigger and stronger than *H. sapiens*. Body proportions were also different with the upper body and arm length being longer in relation to the legs than in humans. Neanderthals also had larger skulls with heavier eyebrow ridges but they had smaller slanting foreheads. Overall, their brains were bigger but were also differently shaped with a smaller frontal area and a flatter brain case protruding somewhat at the back. Preliminary DNA evidence now adds to the general agreement that these morphological differences constitute sufficient grounds to distinguish *H. neanderthalensis* and *H. sapiens* as separate species.

Neanderthals endured considerable extremes of climate. They left abundant evidence of having had plenty of food to eat yet no evidence of structured campsites. They made beautiful tools, known as the Mousterian tool industry, which was based on the *prepared core* technique in which suitable stones were expertly selected and worked. After initial chipping and flaking, a carefully aimed strike would break off a large flake, leaving a sharpened tool nearly ready for use. This invention of this technique cannot be attributed with certainty to Neanderthals but they clearly exploited it. Their ability to work stone points led to skilled hunting of large animals. Nevertheless, Neanderthals did not make more advanced tools or weapons from other materials like bone.

Whether Neanderthals had speech is unclear. Anatomically, it seems plausible but there is a lack of coincidental cultural sophistication or use of symbols that would be anticipated to accompany language. They had heavily built long bones that are consistent with abundant diets and a large capacity for physical exertion. However, some features of their skeletons also suggest they experienced specific nutritional deficiencies (see Chapter 14). Their heavier skeletons have often been assumed to arise from an increased level of exertion but this cannot be the whole story because their children and infants were similarly heavily built. Hence, these contrasts, particularly the scant evidence for basic cultural practices such as burial, make it difficult to attribute the Neanderthal lifestyle, diet, and technological achievements to having developed advanced cognitive function or being able to exploit their large brains.

Homo sapiens

Anatomically modern humans *(H. sapiens)* have existed for at least 100,000 years, with some authorities claiming as long as 300,000 years. *H. sapiens* seems to have come from Africa about 130-160,000 years ago. Nevertheless, clear evidence for a creative explosion does not appear until about 30,000 years ago with the prolific and now well documented cave art that is found especially in France and northern Spain. This was the creative work of the Cro-Magnon who exhibited a major leap forward in brain function over any hominids. Their cave paintings show a good understanding of anatomy, proportion, colour and perspective.

The realistic nature of much of their cave art is all the more astonishing given the extremely difficult working conditions they chose for their art, including inaccessible, dark caves. In some caves they chose to work lying down and worked on the ceilings in cramped spaces only a meter high, and using poor light from candles lasting no more than a couple of hours.

Those whose drawings, paintings and etchings have been found clearly took their art very seriously and, in all likelihood, these artists were respected then at least as much as Renaissance or contemporary artists are today. Art seems to have been central to their lives. The

Cro-Magnon also made rope, fired clay figurines, straightened bone for spear points, made musical instruments, and also provided elaborate burials for some of their dead.

It is reasonably clear that Neanderthals coexisted in time and in several geographical regions with early *H. sapiens*, perhaps even in relative proximity. Whether early humans and Neanderthals interacted and, if so, how, is the subject of much speculation. Indeed, whether the Neanderthals disappeared because early humans killed them off or because they were less adaptable to climatic change is debated. Recent DNA evidence makes it unlikely that Neanderthals were genetically integrated with *H. sapiens*. It is clear that *H. sapiens* is still a relatively young species and is the only *Homo* species present today.

Bipedalism, Smaller Jaws and Bigger Brains

Bipedalism is a strange invention. Its evolution has long been rationalized as being advantageous because it freed the hands of the first hominids from helping with locomotion. Thus the hands could do other things including eating, carrying, tool making, etc. But evidence of bipedalism in hominids precedes evidence of stone tool making by about two million years. If tool making didn't drive bipedalism, what did? The Ohio State University anthropologist, Owen Lovejoy, commented that – *'for any quadruped to get up on its hind legs in order to run is an insane thing to do'*.

Bipedalism largely committed hominids to a ground-based existence. This probably led to a more nomadic existence because, once refined, bipedalism is a more efficient way of walking than the knuckle walking of other primates, which are either tree bound or quadrupedal. Quadrupedalism favours balance and climbing in trees, activities for which humans are ill-designed. Bipedalism might have yielded advantages in hunting animals on the ground but this isn't necessarily so. In fact, the human version of bipedalism has significant disadvantages. Human babies can't walk or even crawl for six months to a year, but the babies of other primates are physically much more mature. How, then, did imperfect bipedalism and immature babies help humans evolve? Only by not being a hindrance to survival.

Moving from *H. habilis* to *H. erectus* and then to *H. neanderthalensis*, the proportion of the overall head size occupied by the jaws shrank and the teeth became smaller. The walking stance became more upright, overall height increased and a larger brain case developed. The implication is that changes in these features are probably related. A smaller jaw and teeth combined with taller bodies probably means a better quality diet and less need for chewing to extract nutrients and energy from food. A bigger brain definitely meant higher energy demands but it also implies potentially better manual, intellectual and social skills. The reasonable conclusion is that hominids having access to a higher quality diet had to chew less and, therefore, had enough energy left over in the diet for fuelling further body growth as well as brain expansion.

Digestion releases the energy and nutrients in food, but some foods are more difficult to digest and yield correspondingly less net energy. Digesting fibrous food releases some absorbable carbohydrates but requires a larger gut and more time, both of which diminish the net nutritional value of fibrous foods. The net energy content of plant foods is therefore lower than from meat or fish. Hominid fossil teeth and jaws started to become smaller at about the same time as the skeleton was adapting to upright stance and bipedalism. Thus, by moving very differently and possessing different teeth and jaws, early hominids undoubtedly ate different food than did quadrupedal primates. Such different food selection would, in turn, probably be dependent on hominids living in a different habitat than that occupied by other primates.

H. sapiens kept up the trend for more erect walking, a more condensed jaw and the larger forehead first introduced by *H. erectus*. However, overall body and skull size is 10-15% less in present day than in early humans (the Cro-Magnon). Brain size is even a few percent smaller than in Neanderthals. If bigger brains means more intelligence and more ability to adapt to or modify the environment (shelter, warm clothing, etc.), Neanderthals would be expected to be the only current *Homo* species because they were brawnier and brainier than *H. sapiens*. However, it was early *H. sapiens* that ousted or at least replaced

Neanderthals. Something about the skills, adaptability, or intelligence of
H. sapiens exceeded that of *H. neanderthalensis.*

Stone Tools

Chimpanzees and a few non-primate species use primitive tools, such as
thin sticks to probe termite colonies, or stones to break nuts or seashells.
Still, for a very long time, humans and certain hominids before them
have been the most sophisticated toolmakers. The usefulness of sharp-
pointed tools as weapons of predation or defense, or to procure food,
seems self-evident and has influenced how human evolution, food
selection and, indeed, human identity are viewed.

However, learning to make and use truly efficient stone tools such as
spears, arrows or knives must have taken a very long time. Some present
day anthropologists have specialized in the making of stone tools (stone
knapping) and have needed many years to learn how to select the correct
types of stone and work them to give a tool equivalent to those made
100,000 years ago. In approaching the problem of how just the right
stone was knapped into an efficient cutting tool, a present-day human has
all of the recorded history of tools and weapons on which to consciously
or unconsciously look back upon.

A present day stone knapper also has the advantage of knowing that a
product of a particular specification could be produced because so many
of them made long ago still exist today. If motivated, modern-day, large-
brained humans take ten or more years to perfect their skill at stone
knapping, it is reasonable to assume that many generations of somewhat
smaller brained early hominids also had to endlessly tinker with and
refine this activity before stone artifacts could be reliably used for any
specific purpose, whether it was hunting with spear points or cutting and
trimming hides. How were they clothing, protecting and feeding
themselves before adequately developing these skills?

Furthermore, many Acheulian stone tools are more refined than
appears necessary for the tasks for which they are assumed to have been
made. This added degree of effort would be like sculpting ornate, in-laid
handles on cutlery used every day: attractive to look at but not really
necessary. Perhaps early *H. erectus* enjoyed the stone-knapping process;

perhaps function is interpreted today when they actually intended art. In any case, when one has enough time that a device is made more carefully and in more detail than its function appears to call for, the maker can be assumed to have had plenty of time on his or her hands. Supper must not have been an urgent issue. Food and leisure time must both have been plentiful so that pursuits other than survival could occupy these hominids.

Regardless of the use to which stone tools might have been put, who was making them? Did hominid clans have families specialized in this craft or did everyone try it, some more successfully than others? Whomever it was, they either had no difficulty feeding themselves from readily available food choices, or else someone was specialized in collecting food while the knappers knapped (or napped). Either way, it means that food had to be abundantly available. The hunting and gathering process must not have been challenging if one could feed oneself or others quite easily.

Art

Like tool making, making easily recognizable images of animals is another highly skilled activity that irreversibly distanced the Cro-Magnon from hominids. Anthropologists and art historians agree that the level of skill (use of perspective, colour, shading, anatomical correctness, and knowledge of seasonal changes in fatness of herd animals) shown in cave art and in the earliest known sculptures is close to that displayed by present day artists who have trained for many years. The important distinction is that artists working in the last few thousand years have mostly had the advantage of working with better materials and under much more amenable circumstances than in caves lit only by primitive and short-lived candles.

As with tool making, the skills to become an accomplished artist would have taken time, even more so if the brain was more primitive than today. Presumably, as now, many more hominids were dabbling in art than were exhibiting their work publicly or were given the honour of painting in the caves. Like with tool making, someone had to be finding food for the artists or it must have been sufficiently abundant that they could easily find enough for themselves without taking too much time from their art.

The point regarding both tool making and art is that (as with most skilled activities) a much larger number of people is usually engaged in the activity than is actually producing the final products of high quality. If, like grazing animals, your days are spent feeding yourself, sleeping, or in self-defense, you stand zero chance of developing the skills of an artist or toolmaker. If, like most primates, you spend only part of your day feeding, sleeping or defending yourself, and have time left over to amuse yourself and socialize, you have at least a slim chance of acquiring the skills to become a tool user or an artist.

Before the advent of agriculture 10,000 years ago, hunting and gathering as they are conventionally described would have been a major preoccupation of virtually every able-bodied individual. Hominids would also have had weekly or seasonal periods of hunger and perhaps famine. Who, then, had time to sit around and 'invent' art or music or stone tools when they were thinking about tomorrow's if not today's supper? Who would have had time to become good enough that the tools actually worked or the art was appreciated enough to be encouraged and allowed on seemingly sacred sites like secluded cave walls? There is no adequate explanation for how these sophisticated skills emerged and were endorsed and refined by the pre-human clans and societies that existed more than 100,000 years ago.

Chapter 2

The Human Brain:
Evolution of Larger Size and Plasticity

Evolution of Neuronal Complexity

The mechanism that makes tissues electrically responsive or excitable
has changed little in the 500 million years since the major animal
families last had a single common ancestor. As first shown over fifty
years ago by Hodgkin and Huxley in the squid giant axon, action
potentials are propagated *'by means of an electrical duet played between
sodium channels and delayed rectifier potassium channels'*, (B Hille,
University of Pennsylvania). When the action potential arrives at the
nerve terminal, calcium channels are used for the rapid release of
neurotransmitter vesicles. How did these excitable systems arise?

Eukaryotic cells (possessing nuclei and other internal organelles)
arose about 1.4 billion years ago. Their emergence involved what must
then have been radical cellular innovations, such as mitochondria for
cellular respiration, as well as the organelles for processing amino acids
into proteins, including the *endoplasmic reticulum, Golgi apparatus,* and
lysosomes. These internal systems of the cell engineered new proteins,
which had not previously existed including *tubulin, actin, myosin,* and
calmodulin. This diversification with new proteins and new membranes
surrounding new organelles permitted the evolution of new cellular
processes, including sorting and packaging of membrane proteins into
vesicles, secretion of the contents of the vesicles, development of an
internal skeleton permitting cells to move and shape change, and more
refined control of cellular activities using calcium as a second messenger.

Multicellular eukaryotes arose in three groups about 700 million years ago – the fungi, the animals and the plants. The principle difference between animals and fungi or plants is the ability to make more rapid movements. By 525 million years ago, the major animal groups were established, i.e. the jellyfish, worms, arthropods and chordates. All nervous systems are based on the same fundamental conditions for membrane excitability that connect sensory and conducting cells, namely the need for cell membranes, ion gradients, gated ion permeability changes, and an effector system capable of responding to the resulting signal.

Neuronal complexity arose through increased *connectivity* between these cells. HJ Karten (University of California, San Diego) describes the evolution of the cerebral cortex as having involved at least two separate and specific steps. First, distinct populations of neurons emerged, each with a characteristic type of interneuronal connection. Second, these distinct neuron populations became laminated, a process occurring as a result of sequential migration to form cell layers from the progenitor cells or *neuroblasts*. Neurons with a single common neuroblast migrate together to form a distinct cell layer within the cortex.

Transmission of a signal across a *synapse* between two neurons can be electric or chemical. Electric transmission between two or more electrically responsive cells requires a bridge or *syncyctium* between the cytoplasm of one cell and another, a feature that can exist in fungi, plants and animals. Only animals have true *gap junctions*, in which the cytoplasm is separate but the membranes of the two cells are essentially fused. Chemical transmission across a synapse requires several features, including neurotransmitter packaging into vesicles within the cytoplasm, controlled release or exocytosis of the vesicle contents at the plasma membrane, and ion channels sensitive to the chemical neurotransmitter.

Chemical neurotransmitters are common molecules. Their uniqueness in relation to signal transmission comes in positioning their release sites on the pre-synaptic cell near the receptors for these chemicals on the post-synaptic cell, and in the speed and modulation of their calcium-mediated response. Bacteria are *prokaryotes* that can synthesize several neurotransmitter molecules but do not meet the other criteria for chemical transmission, which are present in all eukaryotes.

Protein Homology

The relevant proteins used as enzymes, ion channels, and receptors in these neuronal communication processes are very similar between yeast and mammals so they have evolved little over the past billion years. Modern protein evolution involves mosaics formed by the assembly of multiple protein blueprints that were repeatedly recombined on a common template, resulting in complex proteins with specific domains for substrate recognition, catalysis, modulation and protein-protein recognition.

Homeobox or *'hox'* genes are central to expression of specific characteristics in developing brain and many of these genes are homologous across extremely diverse species, such as humans and the housefly, *Drosophila*. The bacterium, *E. coli,* has sodium-potassium pumps that, although not used for signal transmission, are homologous to those in plants, yeast and animals. This commonality implies that the basic components needed to evolve complex brains arose only once. Fundamentally different forms of respiration, locomotion and sexual reproduction exist but complexity in the nervous system follows a recognizably similar pattern across diverse animal species.

Karten proposes that the nervous system's dependence on a core blueprint arose with the evolution of homologous proteins in widely differing species and with the homologous domains on single proteins that are functionally linked despite being used for diverse purposes. Such homologous domains include cyclic nucleotide phosphorylation, cyclic nucleotide ion channel gating, voltage gating of cation channels, and membrane pore-forming regions.

The structure of some of the many different voltage-gated potassium channels in the mouse is more similar to these channels in the housefly than to other potassium channels in the mouse. These protein homologies exist within animals but not in plants or fungi, suggesting they were a fundamental part of the evolution of nervous systems. Calcium channels probably evolved from potassium channels and sodium channels probably evolved from calcium channels before the wide diversity in potassium channels evolved. Sodium channels make rapid conducting axons possible, which allows signals to be transmitted without unnecessary and excessive use of calcium-dependent messaging during each stimulus.

The earlier evolution of mechanisms to maintain potassium and calcium gradients suggests they became more important than sodium or chloride gradients because many bacteria can grow in the complete absence of the sodium or chloride ions. Rapid conducting axons using sodium channels would only be necessary in large animals, which evolved later than smaller multicellular organisms. The evolution of these three types of ion channels allowed for specialization of some cells specifically for signal conduction rather than for contractile and secretory functions.

The Brain in Primate Embryos

The large difference between the brain size of humans and other primates is almost nonexistent in early stage primate embryos, such as humans and macaques, in both of which the brains occupy about *one third* of overall body size. Not only are the brain sizes relatively much larger than at any later time in development, but they are also virtually the same size at the same stage of development in these two very different primate species (Figure 2.1). This striking evidence is uncommon because these comparisons are not often reported. Nonetheless, this example suggests that if, as embryos, primates as different as humans and macaques have similarly large brains, then humans have no specific anatomical or genetic advantage for brain growth that shows up early in embryonic development. Rather, the developmental differences that favour much larger brains in humans really emerge later in fetal development and continue to widen as development proceeds postnatally to adulthood.

A simple anatomical comparison such as similar brain weight in embryos of humans and macaques is inadequate proof that there are really only minimal differences in the genetic potential for brain growth between humans and other primates. Genetic differences between species show up in the way and rate that a species develops all the way from embryo to newborn to adult. However, differences in the genetic potential for primate brain growth may well remain small until the fetus gets to the stage at which the brain starts to make significant metabolic

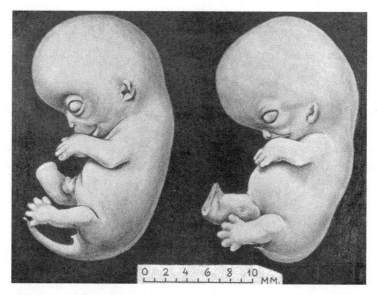

Figure 2.1. Similar and large brain to body proportions in a macaque (left) and human (right) embryo, both aged about 45 days. Modified from Schultz (1969).

demands, i.e. the stage at which it has sufficient working cells that its oxygen and energy demands start to mount. Genetic differences in the ability to develop a blood supply to the brain and to provide it with energy and nutrients will then start to play a role in influencing the duration of neuronal network expansion, which is a key component of brain growth and enlargement.

Extraordinary Connectivity

The attributes of consciousness, imagination, problem solving, adaptability to new situations, and language (both spoken and written) are important components of advanced cognitive function in humans. They are dependent on attaining a more sophisticated level of information processing than is present in any other land-based species besides humans. Large brain size relative to body size appears necessary for advanced cognition but the threshold of brain size and organization at which intelligent thought becomes possible is still unclear.

Viewed in terms of evolution, the question is how the prerequisite of advanced cognition - *extraordinary connectivity* - amongst neurons in the human brain got started in the first place. The fly in the ointment is that the presumed stimuli for advanced cognition - language, culture and organized social activity - all already require an advanced brain. If one already possesses an advanced brain, one can learn new skills and can improve them with practice. But are rudimentary efforts to acquire new skills passed on the next generation? Does a well-trained chimpanzee produce smarter offspring? Humans speak because we have advanced brains and the appropriate anatomy and neural circuitry. Language acquisition can be improved by interacting with family and others in society. But which comes first, the chicken or the egg: social, intellectual and manual proficiency or a bigger more advanced brain?

This is a difficult but important issue in explaining human brain evolution because culture, hunting, and tool making are all to a greater or lesser extent seen as the foundation of the larger and more advanced human brain. However, this is a circular argument because a more advanced brain is needed before one can improve the proficiency of neuronal activity commensurate with intelligent thought. How did humans get bigger, more sophisticated brains without the skills that already need a bigger brain?

The usual answer is that it happened incrementally – a little at a time. Hominid societies became larger and the intricacies of culture started to emerge. People in close proximity start to learn skills from each other. Larger social groups, i.e. villages, require a stable habitat and food supply but don't require bigger brains. That is reasonable but it really does have to start somewhere. How did hominids start to congregate in larger groups than other primates? There aren't dozens of large-brained, cognitively advanced primates on which to model human brain evolution; there is only one.

Something had to increasingly propel certain hominids into increasingly social interaction and it had to be something that had no real or imaginary benefit for their survival; natural selection governs evolution and has no vested interest in survival or the emergence of any particular characteristics, including bigger brain or higher cognitive function.

Survival Versus Plasticity

The misleading thing about the human cognitive capacity is that the benefits receive most of the attention. Human intelligence is unsurpassed in its potential to solve problems and to do beautiful and imaginative things. But this cleverness has its down side too. In order to achieve moderate to high intelligence, humans need the correct genetic endowment. Minimum standards of nutrition and psychosocial development also need to be met in the first few years of life. Furthermore, achieving reasonably intelligent and constructive behaviour in later life depends on graduating through a series of *critical windows*.

Cruelty, neglect, malnutrition or extreme poverty all exert a negative influence on human brain and psychological development; they may not prevent achievement of high intelligence but they can permanently distort one's perception of the world and greatly reduce the odds of normal mental development. Thus the psychosocial and nutritional setting during the critical windows are enormously important for later attitudes.

The existence of critical windows during infancy and childhood means that normal cognitive development is not assured, i.e. it is vulnerable. This important issue is dealt with in more detail in Chapter 4. The emergence of the high functioning human brain demonstrates that this developmental vulnerability can usually be avoided but this doesn't mean the vulnerability wasn't a problem as brain expansion began, nor does it mean that it has disappeared. Iodine deficiency disorders and cretinism in over 700 million people worldwide today makes this very clear (Chapter 6). The point is that hominid brain evolution had somehow to deal with and accommodate to this evolving vulnerability.

Added freedom of choice and the ability to decide between different options lead to the potential for greater extremes of behaviour in humans than in other species. Humans are capable of worshipping other people or killing them or themselves. Humans can choose to copulate with one or numerous partners, heterosexually or homosexually, or with none at all. Humans are capable of lethal self-starvation in the face of plenty (anorexia) or of overeating to the point of incapacitating obesity.

Humans can devise intricate and beautiful objects or can be wantonly destructive, risking the destruction of humanity.

As David Horrobin brilliantly described in *The Madness of Adam and Eve*, the same brain that is capable of creating beautiful art, music, architecture or anything else that most people would admire is, through disease and psychosis, also capable of a highly distorted view of self and of the motives of others. As well as harbouring immense creativity, the humans can be deceitful, cruel, paranoid and irresponsible. The neurological basis of creativity clearly involves a dark side as well. In evolutionary terms, the propagation of madness and extreme anti-social behaviour for the sake of stone tools seems to be a risky venture.

In *Promethean Fire: Reflections on the Origin of Mind*, Charles Lumsden and Edward Wilson note that despite its distinctive and powerful capacities, the human mind can also be remarkably *inefficient* in some ways. They note that humans are relatively poor at information recall, are unable to adequately judge the merits of other people, and are poor at estimating risk. As well, they make excessive use of stereotypes while planning strategy. Lumsden and Wilson claim that humans tend to equate events that have low probability and trivial consequence with events that have low probability but important consequence. Therefore the lasting impact of catastrophes, such as warfare, floods, windstorms, droughts, etc., is generally underestimated even when the same catastrophe is repeated and experienced by several sequential generations.

Compared to whales or birds that communicate with long, complex sound sequences, humans also have a poor ability to faithfully replicate sound or movement. This includes poor memory for numerical sequences like the arithmetic term 'pi' to multiple decimal places. This must have been an impediment to acquiring speech. *Idiots savant* may excel at this rote-type capacity but, paradoxically, are often otherwise intellectually subnormal and socially dysfunctional. Humans also have poor memory for space and time compared to, say, honeybees, which can faultlessly traverse a complicated route from a new food source and accurately report its location to members of the hive.

There is also a tendency to equate today's more advanced technological achievements with higher intelligence. The possibilities offered by technology whether it be exploration of Mars, high efficiency automobiles, or video-cell phones are impressive and, despite their desirability, were not possible even fifty years ago. Does that make today's youth inherently smarter than their grandparents or people living in the Middle Ages?

Present day technology builds on previous technology and information but, in lamenting the decline in wisdom during his time, T.S. Elliot noted that knowledge does not substitute for wisdom nor information for knowledge. And that was in 1934! We certainly have more information today than ever before but it is not clear that any more information filters through to become wisdom than it did a century or even a millenium ago. Thus, despite disproportionately large brains, humans do not uniformly excel in all aspects of neurological activity or decision-making. Indeed, we are much worse at some activities than some rather humble insects.

The point is that human intelligence and normative behaviour are both *plastic* – there is no rigid blueprint for either. That is precisely the point. To function normally, termite and bee colonies and, to a lesser extent, even mammals require and receive specific blueprints for behaviour and survival. Humans do not and as a result outcomes related to survival can potentially be a lot better but also potentially a lot worse. Brain expansion and advancement during human evolution was thus a double-edged sword.

Brain Weight, Body Weight and Encephalization Quotient

Despite today's modern imaging technology and accurate balances, there is still discrepancy as to the average normal adult human brain weight. Part of the discrepancy arises from differences in what is being measured. At what anatomical landmark is it no longer the brain and becomes the spinal cord? Is volume or weight being used? Is it the brain of a 20 year old or one of an 80 year old, because aging alone causes the brain to shrink. Is it the space inside the empty dry skull or the brain itself, stripped of all protective membranes? The volume of the brain's cavities and *cerebrospinal fluid* they contain vary considerably.

John Allen and colleagues at Washington University in St. Louis have recently reconstructed human brain volumes from magnetic resonance images. They reported that the 'empty' brain of young, male, adult humans has a volume of 1273 cubic centimeters (cm^3), a value that increases to 1360 cm^3 if the associated fluids but no surrounding membranes are included. The reference value for modern human brain size that I will use here is therefore 1360 grams, which is equivalent to 1360 cm^3. Allen's group confirmed that male brains are generally larger than female brains by 10-15% in all regions. They also showed that, compared to male brains and to other areas of the female brain, the left occipital (back) region of the female brain has the most variable volume.

Allen's images confirm many previous reports showing that the occipital lobe has undergone extensive reorganization in humans compared to other primates. In contrast, the visual cortex is larger and spread more laterally in non-human primates. Allen's values are consistent with other imaging studies and provide a useful basis for understanding why some of the variability occurs between other reports.

Brain weight is widely claimed to increase proportionally (allometrically) with body weight but not quite as fast as the increase in body weight. However, Michael Crawford at London's Institute for Brain Chemistry and Human Nutrition has pointed out for many years that, in fact, the brain shrinks in proportion to increasing body size, something that is evident from the very small brains and large bodies of the big savannah herbivores like the rhinoceros (Table 2.1).

Humans have larger brains than expected for their body size, whether compared to other primates or to hominids. To have a measure of this 'overencephalisation' in humans, Harry Jerison from the University of California at Los Angeles, introduced the concept of *encephalization quotient* (EQ). Published EQ values differ somewhat according to what adjustments are to Jerison's formula but, in moving from Australopithecines through more advanced hominids to living humans, all studies show EQs increasing two to three fold (Table 2.2). That means that humans have two to three times more 'corrected brain capacity' than in their forebears.

Table 2.1. Average body weights (kg), brain weights (g) and brain to body weight ratios (%) in a range of small, medium, large and very large mammals. Note the marked dichotomy between brain weights of the large marine mammals versus the large terrestrial mammals[1].

Species	Body Weight	Brain Weight	Brain/Body Weight
Tree Shrew	0.3	4	1.3
Marmoset Monkey	0.4	13	3.3
Grey Squirrel	0.6	20	3.0
Cebus Monkey	1.9	63	3.3
Cynomologus Monkey	3.0	110	3.1
Chimpanzee	45	400	0.9
Human	60	1360	2.3
Porpoise	125	1400	1.1
Bottle-nosed Dolphin	165	1600	1.0
Gorilla	170	500	0.3
Ox	500	450	0.10
Black Rhinoceros	1200	500	0.04
Elephant	5000	7500	0.15
Sperm Whale	37000	7820	0.02

[1]Modified from Cunnane *et al* (1993).

The EQs shown in Table 2.2 are not the same as those in the published literature because I have arbitrarily given *H. sapiens* a relative EQ of 100 and then took EQs of hominids and living apes from the literature and expressed them as a percent of the human EQ. This doesn't give humans even more braininess, but simply converts all the numbers to a percent scale, using one species EQ as a reference.

The average adult human brain weighs 1360 grams (about 3 pounds) and the average human body weighs 60 kilograms (or 132 pounds; Table 2.2). The early hominids, *A. afarensis, A. garhi,* and *A. africanus,* had brains about one third the weight of human brains. *A. afarensis* had a body weight about half that of humans but other hominids such as *A. africanus* and the *Paranthropus* species had lean body weights that come closer to those of humans. Thus, *A. afarensis* had a brain to body weight ratio of about 1.7% (455 grams divided by a body weight estimated to be 27 kilograms).

Survival of the Fattest

Table 2.2. Brain and body dimensions of extinct and extant adult hominids, humans and related primates.

Species (time period)	Brain Weight[1] (grams)	Body Weight[1] (kilograms)	Brain/Body Weight (%)	Relative EQ[2]
A. afarensis (3.6-2.8 mya)	455	27	1.7	41
A. garhi (2.5 mya)	450			
A. africanus (3.0-2.2 mya)	445	46	1.0	44
P. aethiopicus (2.7-2.3 mya)	405	38	1.1	44
P. boisei (2.3-1.3 mya)	500	53	0.9	46
P. robustus (1.8-1.5 mya)	520	47	1.1	50
H. rudolfensis (2.4-1.7 mya)	750	46	1.7	59
H. habilis (1.9-1.5 mya)	650	35	1.7	57
H. ergaster (1.8-1.4 mya)	855			
H. erectus (early; 1.8-1.5 mya)	863	60	1.6	63
H. erectus (late; 0.5-0.3 mya)	980	58	1.6	63
H. heidelbergensis (600-200 kya)	1200	66	1.8	74
H. neanderthalensis (200-40 kya)	1420	76	1.9	75
H. sapiens (early; 100-10 kya)	1490	63	2.4	102
H. sapiens (adult male)	1360	60	2.3	100
(corrected to lean body weight)		51	2.7	
H. sapiens (newborn)	400	3.5	11.4	
(corrected to lean body weight)		2.9	13.3	
Pan troglodytes (adult male)	400	45	0.9	42
Pan troglodytes (newborn)	180	1.6	11.3	
Pongo pygmaeus (adult male)	400	80	0.5	32
Gorilla Gorilla (adult male)	500	170	0.3	25

[1]These data are averaged from publications by Aeillo and Dean (1990), Pinker (1997), Ruff et al (1997), Kappelman (1996), MacKinnon (1978), Falk (1987), Cronin et al (1981), Passingham (1985), Blumenberg (1983), and Leonard and Robertson (1994), Schultz (1969).

[2]Encephalization Quotient: Data are averaged from Ruff et al (1997), Martin (1981), and Kappelman (1996). For easier comparison, they are standardized to a new term - relative EQ – in which living *H. sapiens* are set at 100 and comparisons become percentages of the human value.

mya/kya – million/thousand years ago.

A brain to body weight ratio of 1.7% is about three quarters of the 2.3% value in modern humans (1360 grams divided by 60 kilograms). Hence, the three fold absolute difference in brain weight between humans and *A. afarensis* is greater than the difference in their brain to body weight ratios because the body weights are not as different as the brain weights. Using my relative EQ of 100 in *H. sapiens* as a reference, *A. afarensis* had a relative EQ of 41, meaning that their corrected brain capacity was about 41% of that in humans today. Because humans have evolved larger bodies as well as larger brains, current human brain capacity is more similar to that in early hominids than the three fold difference in brain size alone would imply.

Brain size and relative EQ both increased by 40-45% in going from *A. afarensis* to *H. habilis* (Table 2.2). Similarly, in going from *H. habilis* to *H. erectus*, there was a further 40% increase in brain size but relative EQ increased much less because *H. erectus* was considerably bigger than *H. habilis*. *H. heidelbergensis* had a 20-30% increase in absolute brain size over *H. erectus* but, again, body weight increased so the change in relative EQ (74 versus 63) was less than in absolute brain weight. Neanderthals gained both brain and body weight so they had almost identical brain capacity to the somewhat earlier *H. heidelbergensis*.

Early humans (Cro-Magnon) had the largest known hominid brain size at 1490 grams and had somewhat smaller bodies than Neanderthals so they experienced a relatively large (36%) increase in EQ compared to Neanderthals. Compared to the Cro-Magnon, living humans have lost about 130 grams of brain weight but have also lost about 3 kilograms of body weight, resulting in both the brain to body weight ratio and relative EQ of humans slipping 2-3% over the past 20-30,000 years.

These changes in relative brain size between hominids and humans can be condensed into four broad phases (Table 2.3): First there was about a 40% increase in EQ over the 1.5 million year period between *A. afarensis* and *H. habilis*. Second, there was only a small increase in EQ between *H. habilis and H. erectus* because body growth paralleled brain expansion. Third, there was about a 62% increase in brain size and EQ between *H. erectus* and *H. sapiens*, because brain growth again considerably exceeded body growth. Finally, there was a small, parallel reduction in both brain and body size in the past 30,000 years that slightly reduced EQ in present day compared to early humans.

Survival of the Fattest

Table 2.3. The four broad phases of human brain evolution.

	Approximate Duration	Species Involved	Main Change in Brain Size
Phase 1:	1.5 million years	*A. afarensis* to *H. habilis*	40% increase in EQ because brain growth exceeded body growth
Phase 2:	1 million years	*H. habilis* to *H. erectus*	Minimal change in EQ because body and brain growth occurred in parallel.
Phase 3:	400,000 years	*H. erectus* to early *H. sapiens*	62% increase in EQ because brain growth exceeded body growth.
Phase 4:	30,000 years	Early to present *H. sapiens*	Small parallel reduction in both body and brain weight.

These four phases support three important concepts in human brain evolution: First, sporadic change with periods of slow to negligible growth occurred, rather than continuous, gradual change in brain and body weight. Second, changes in brain and body size did not necessarily occur at the same time. Third, changes in brain size (upward or downward) always seem to have accompanied upward or downward changes in body size but increased brain size occurred twice when body size seems to have changed very little (first as Australopithecine became *H. habilis* and again as *H. erectus* evolved to *H. sapiens*).

Table 2.2 also includes EQs for living non-human primates. Lean body weights of humans are almost the same as in chimpanzees, so this comparison shows a two to three fold advantage for humans whether absolute brain weight, relative brain weight or EQ is considered. Adult gorillas are nearly three times bigger than human adults but still have brains the size of early hominids (both 450-500 grams). Unlike in hominids, impressive body growth can clearly occur in other primates without concomitant brain growth.

A Milligram Every Year

Body weight was essentially the same in *H. erectus* and *H. sapiens*, so the rate of change in brain size between them can be discussed without correcting for differences in body weight or EQ. Brain size increased by about 500 grams over the 400,000 years between later *H. erectus* and early *H. sapiens* (Table 2.2). An increase of 500 grams over 400,000 years is equivalent to a one milligram increase per year. A change in brain size of one milligram every year seems trivial, especially when a difference in brain size within a single human family of a hundred or more *grams* is completely normal today.

Some periods (thousands or perhaps tens of thousands of years long) probably saw more expansion than did others. For instance, between 200,000 and 40,000 years ago is generally regarded as the fastest period of brain expansion, whereas the million years between Australopithecines and early *H. erectus* is generally regarded as the slow period (although body growth was substantial during this earlier period). Fast or slow, ultimately, there were about four million years of cumulative brain expansion.

Then, suddenly, the four million year old freight train of pre-human brain expansion ground to a halt. Average brain size stopped increasing roughly 30,000 years ago, which, paradoxically, was not long after the time when hominid fossils became close enough to those of modern day humans to be classified as truly human. In fact, the freight train has been slowly reversing since then because, compared to the earliest 'modern' *H. sapiens* who had brains weighing an average of 1490 grams, the average size of human adult brains is 1360 grams, or 130 grams smaller today. That's a loss of slightly more than four milligrams of brain per year, which is a rate of loss of brain size *four times* faster than the rate of gain made over the preceding 500,000 years.

The large number of skull samples of both Cro-Magnon and present day humans precludes miscalculating their average brain size to yield a loss of 130 grams over 30,000 years. Perhaps the loss of brain size that humans are currently experiencing is a temporary dip in what would otherwise be truly considered a plateau. A time period of 30,000 years is really too short to know for sure. If real, such a loss in human brain size has implications for mechanisms of its evolution.

Incidentally, disregarding the threefold increase in brain size from infancy to adulthood, Lars Svennerholm at Gothenberg University has shown that, after about 25 years old, the human brain actually shrinks by about 8% or just over 100 grams over the next 50-60 years. In fact, the normal shrinkage humans experience during a lifetime equals or slightly exceeds the loss of brain size our species has experienced over the past 30,000 years. Some diseases like autism are associated with somewhat larger brain size while low birth weight and prematurity commonly retard brain growth and development.

Challenges to Interpreting Hominid Fossil Crania

Few paleontologists or anatomists have the knowledge of hominid fossils of Philip Tobias from Witwatersrand University in Johannesburg, South Africa. Tobias has emphasized that there are many challenges to interpreting hominid fossil crania, starting with the paramount need for careful dating and good identification of skulls or *endocasts* (casts made of the inside of skulls). In his opinion, all that can really be objectively evaluated in fossil skulls is size, shape and surface impression. In the first 2 to 3 million years of hominid evolution, these criteria limit discussion to only a few good specimens. Even in the reliable specimens, the paucity of good fossils and the variation in shape and size leads to disagreement over the actual size of the brain.

Body weight in hominids is even more uncertain than brain shape or volume because it has to be estimated from one or other feature of the skeleton believed to relate directly to body weight, i.e. the length of leg bones or width of the pelvis, etc. Hence, an error factor is built into all such measurements. Again, there are frequently only a few specimens to start with and there isn't good agreement about which skeletal features best correlate with body weight.

Despite being of similar absolute size, Australopithecine cranial endocasts have several clear attributes distinguishing them from the brains of the great apes: First, in Australopithecines, the hole in the base of the skull through which the spinal column passes (*foramen magnum*) is located forward of its position in apes. This places the spinal column of Australopithecines more directly under the skull rather than projecting at an angle backward from it as in the apes. This, in turn, is an important feature suggestive of more upright stance if not full bipedalism in Australopithecines.

Second, the parietal lobe is better developed in Australopithecines than in living apes. Third, in Australopithecines and *Homo*, the cerebrum is the most posterior component of the brain so, with rare exception, it occupies the whole top of the head and the cerebellum is centered *under* the cerebrum. In apes, the cerebellum projects more posteriorly (see Figure 2.2).

Somewhat in contrast to Tobias' conservative view, Ralph Holloway of Columbia University in New York sees several important organizational changes occurring in Australopithecines before the brain of *H. habilis* emerged. First, he suggests that 3-4 million years ago, the primary visual cortex at the back of the brain started to decrease in size by as much as 20%. Simultaneous with the reduction in the visual cortex, the *association area* of the parietal cortex at the side of the brain started to increase in area. Holloway proposes that these two changes pushed the significant groove dividing the back from the side of the brain (the *lunate sulcus*) backwards in hominids compared to where it is in all of the apes (Figure 2.2).

Since the parietal association area is used for multi-channel processing of visual, auditory, and somatic sensory information, as well as for social communication, Holloway suggests that the larger association area of the later hominid parietal cortex helped mediate improvements in understanding spatial relationships. Its expansion would appear to be an important reason why language developed in *Homo* but not earlier hominids. In essence, he suggests that the early expansion of the parietal association area is intimately linked to the emergence of truly human neurological capabilities. Holloway sees this improved function as supporting adaptation to harsher savannah life, tool making and bipedalism.

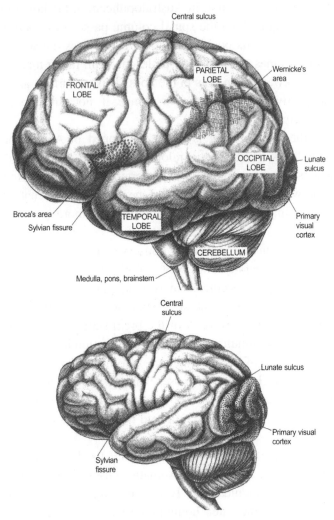

Figure 2.2. Comparison of the morphology of the human and chimpanzee brain (modified from Conroy, 1997).

Two to three million years ago cerebral asymmetries became evident including left occipital and right frontal enlargement. These asymmetries suggest cerebral specialization that may have begun in *A. africanus* but are certainly clear in *H. habilis*. Not long after the reduction in the visual cortex and expansion in the parietal cortex, the frontal lobes started to reorganize. Thus, speech-processing capabilities started to develop in Broca's area 2.5 to 1.8 million years ago. This particular change is first clearly evident in *H. habilis*. Apart from these specific (and controversial) changes, Holloway notes that brain enlargement was proportional to the increase in body size that was occurring simultaneously in other *Homo*.

Vascular Evolution in Hominids

Changes in blood flow in the hominid cranium contribute to understanding changes in brain size and morphology. Primate brains have an inordinately high metabolic requirement for their size so even bigger primate brains require more oxygen and produce more carbon dioxide and wastes. Brain function is also compromised by a rise in its temperature. Since bigger brains produce more heat, they require a commensurate increase in blood supply not only to nourish them but also to dissipate the heat produced during their activity.

The major vessels, especially the *venous sinuses* and *meningeal vessels* that collect the blood between the surface of the brain and the skull, make impressions in the skull. The pulse of the brain's major arteries that lie under the veins and sinuses contributes to the grooved impressions in the skull. Hence, in order to better understand how the hominid brain evolved, it is possible to study changes in cranial vasculature simultaneously with changes in brain shape and size.

The brain's blood drains through rigid-walled venous sinuses of the *dura mater*, which are the protective membranes surrounding the brain. These sinuses are on the sides and top of the brain and connect to the veins, which, via multiple branches, drain into the *internal* and *external jugular veins* in the neck. The dura mater sinuses and the temporal and parietal *superficial vessels* deeply groove the cranium from before birth. The anterior superficial vessels do not groove the cranium since they run lengthwise in the space between the brain's two hemispheres.

The dura mater sinus system is present in the same form from Australopithecines onwards but the meningeal venous system *per se* becomes progressively more complex. Prior to about two million years ago, the middle meningeal vessels were very simple or were not seen at all. *A. robustus* but especially *A. gracilis* are good examples that reflect the simple cranial venous topography of the human neonate. In later hominids, the brain became larger and shows more complexity of venous drainage beginning with branching of the middle meningeal veins.

Despite a much larger brain, Neanderthals still had an old-fashioned meningeal vessel system with poorly developed branching and a large anterior vein on both sides of the brain that does not correspond to drainage features in humans. This simplicity in the Neanderthal brain's venous drainage is especially evident in Neanderthal infants. In contrast to Neanderthals, the earliest modern humans had more branches between the middle meningeal vessels. In these specimens, the most extensive venous branching seems to be at the top of the brain. Full complexity of the middle meningeal venous network seems to have been achieved by 30,000 years ago.

Increased branching and complexity of the brain's venous system must parallel similar changes in the arterial system, but the latter are not traceable because, unlike the veins and sinuses, the arteries are not close to the surface of the skull. These vascular changes must also probably reflect changes in metabolic rate, energy demand and heat production in the brain areas they serve, i.e. in the frontal and temporal cortices where speech was first emerging and where cognitive processing was becoming more sophisticated.

Bipedalism in hominids probably increased venous drainage pressure on the vertebral venous plexus at the top of the spinal column. Dean Falk of Florida State University suggests that this additional pressure and the need for improved cooling via veins on the brain's surface improved venous drainage from the brain in early Homo, which is thought to have helped release a thermal constraint on brain expansion. This is the basis for the *Radiator Theory* of human brain evolution, i.e. once the brain could get rid of more heat through better drainage, its expansion could occur or continue.

The Cranial Base – Is There a Link to Brain Evolution?

The cranial base below the front part of the skull is the interface between the skull and facial bones. Fred Spoor and colleagues of the University

College London explored whether the morphology of the cranial base of human infants could provide insight into the origins of the modern form of the human skull and hence human brain evolution.

In contrast to other primates, several aspects of the cranial base in humans are different: First, the bones cementing the bottom of skull to the roof of the mouth (the *petrous pyramids*) are more upright in humans. Second, the foramen magnum is more tucked in under the middle of the skull in humans. Third, the base of the human cranium is more flexed in midline, meaning that at the front it is more horizontal but, behind the foramen magnum, it becomes angled almost vertically downwards. This makes the depressions on both sides of the back of the lower skull (the *cranial fossae*) considerably deeper and wider in humans than in other primates.

According to the *Spatial Packing Hypothesis,* the shape of the cranial base is seen as responding to the development of the brain and face. In species in which the brain is larger, this hypothesis proposes that brain enlargement is possible due to more flexion of the cranial base, i.e. the brain enlarges in part by expanding backwards but also downwards. Since the human fetal skull grows more slowly at the base than elsewhere, the occipital regions seem to flex downwards under the pressure of this growth because the brain itself, especially the cerebellum, is growing rapidly during the third trimester.

Spoor's group made magnetic resonance images of formalin-fixed human fetuses and found that downward flexion of the cranial base increased by 9° between 10 to 29 weeks of gestation. This flexion occurred in several phases and was greatest at birth compared to either mid-gestation or adulthood. The reduced angle in adults compared to newborns contradicted their expectation that flexion of the cranial base would relate directly to the degree of brain enlargement. Indeed, the size of neither the cerebellum nor the cerebrum was statistically correlated to the amount of flexion of the cranial base or to the final orientation of the petrous pyramids joining the skull to the roof of the mouth. Since cranial base angle in human infants was not at all correlated with final brain size, these data did not support the spatial packing hypothesis. Therefore, increased brain size must be accommodated elsewhere than at the base of the cranium.

Indeed, Spoor's group found that there is actually less flexion of the cranial base in humans than would be expected for our brain size, and suggested that perhaps there are limits to the flexibility of the cranial base beyond which other functions involving the neck, including respiration, might be compromised. They suggest that mechanisms controlling cranial base morphology are different in humans than in non-human primates, with different growth of organs in the neck probably being implicated in the particular ontogeny of the human cranial base. Since speech involves the pharynx, the different cranial base morphology in humans may be directly or indirectly related to the evolution of speech.

Body Fatness: A Key Factor in Relative Brain Weight

When comparing body weights, it is assumed that humans, chimpanzees and extinct hominids all had similar body *composition*. This isn't so because primates normally have very low fatness, especially in their babies. This may seem like an implausible oversight because we are so conscious today of excess body weight caused by fat. Nevertheless, body fatness is never actually taken into consideration in body weight comparisons between humans and other primates or hominids. Correcting for body fat not found in other species actually sets relative brain weights in modern humans further apart from other animals than is commonly recognized.

In healthy adult humans, body fat usually contributes to 15-20% of overall weight. In affluent societies, body fat averages more than 20% of body weight but it is rare that adults anywhere are healthy with less than 10% body fat. However, few other land animals in the wild have more than 5% body fat. Of the fat that can be measured chemically, almost none in other animals is visible as *subcutaneous fat* (under the skin). This difference in body fatness between humans and non-human primates leads to underestimating relative brain weights in humans because they typically have four times more body fat than adult non-human primates.

To compare like with like, not only body weight but also body composition needs to be measured and should be as similar as possible across species. In other words, about 15% of average adult human body weight should be removed so that normally lean non-human species can be compared to the corrected lean weight of humans. This would leave the average lean adult human weighing about 51 kg instead of 60 kg. Correcting for fatness doesn't change actual brain size but it does raise the relative brain weight of modern humans to 2.7% from 2.3% and relative EQ would correspondingly rise to about 117.

Relative Brain Size in Fat Babies

Brain weight corrected to lean body weight is more important when comparing infant rather than adult humans with chimpanzees. It is in infants that the brain is vulnerable and developing infants have relatively much larger brains than do adults. Equally importantly, human newborn babies are positively obese compared to babies of other species, including chimpanzees. The modern day, full term, healthy human baby has an average brain weight of 400 grams and an average body weight of 3500 grams. That makes its relative brain weight 11.4%.

However, about 500 grams or about 14% of its body weight is fat. Hence, an average healthy newborn human has a *lean* body weight of about 3000 grams and has a brain weight relative to lean body mass of 13.3%, which is 13% more than before correcting for fatness. Newborn chimpanzee infants are also lean and have relative brain weights of about 11%, which is the same as in human infants. Removing body fat from the calculation means lean human babies then have similar body composition but have true relative brain weights about 18% larger than in chimpanzees.

Several points arise from these comparisons (Table 2.4): First, newborn human infants have brains that are about one third of the absolute size of those in adult humans. However, this difference is much smaller than for their body weights, those of infants being 1/20 the size of adults. As a result, relatively speaking, the brain weighs *much more* at birth than in the adult. This is true across species, not just in humans, and is an important factor that needs to be considered in discussing brain evolution.

Table 2.4. Key points in the comparison of adult and infant brain size and brain/body weight ratio.

---

1. Body fat is a much larger component of the body weight of healthy human infants (about 14%) compared to other large terrestrial mammals, including *all* other primates.
2. To compare like with like, body fat should be removed from body weight, giving 'lean' human infants a brain/body weight ratio of about 13%.
3. Even without removing the effect of body fat in human infants, brain/body weight ratios are much more similar in human and chimpanzee infants than in the respective adults.
4. Notwithstanding a bigger increase in brain weight as humans mature, the main difference in brain weight between adult humans and chimpanzees is greater body growth in chimpanzees.
5. Body fatness is crucial to more than the brain/body weight ratio in humans; because body fatness after birth is a function of early nutrition, its presence is one of the most important indicators of normal human brain development.

Second, at birth, the difference between the relative brain weights of chimpanzees and humans is smaller than it is in adults. The human infant brain is over twice as big as in chimpanzee infants while in human adults, the brain is about three times bigger than in adult chimpanzees. Although humans and chimpanzees are born with a relatively small difference in brain size, after birth, chimpanzees put more effort into body growth while humans keep putting it into brain as well as body growth.

Third, would it make any difference to their brain development if human babies were born leaner and were, in fact, more like chimpanzees? Yes, is the short and unequivocal answer; it makes a huge difference. Fatness makes a newborn human baby healthy and is essential for normal brain growth and neurological development. Leanness in a human baby is not the normal or ideal situation. Leanness can be healthy in adults but it isn't in babies. Leanness in babies arises due to being born early (premature) or by being undernourished during pregnancy. Either way, in general, lean babies have a much higher risk of slower brain development (see Chapters 4 and 5).

Evolutionary processes are often assessed by comparing the adults of different species. Across primates, the developmental difference in brain size (actual or relative) is greatest in adults so it is understandable to use the adults as the main models. However, the foregoing analysis of relative brain size in human and chimpanzee babies is intended to lay the foundation for the case that it is differences and similarities amongst *neonates* that may be more likely to inform us about human brain evolution rather than the differences between adults.

***

I suggested earlier in this chapter that the small reduction in brain size as early humans (the Cro-Magnon) evolved into present day humans should give us pause for thought. Did the functional capacity of the brain also decline with that loss of size, or was it unaffected? Could genes that promoted increasing brain size during the previous million or more years change and reverse the increase in size?

A facultative parameter like the environment (climate or diet) can wax or wane; that is the nature of environmental events or change. Genetic information is more of a constitutive parameter. It does change, but in this case the reversing trend in brain size implies that the mutations that helped program the increase in brain size got switched off. Actually, it is more than that; if human brain size is really shrinking, the genetic change favouring expansion would have to be more than switched off; switching off would potentially arrest brain expansion but for the brain to shrink, genetic effects on expansion would have to start reversing themselves. One therefore would have to postulate new genetic events that exactly nullified the initial genetic change.

Alternatively, are there indeed environmental changes that have adversely affected human brain size in the past 30,000 years? Agriculture was a major invention and has been widely adopted in the past 5,000 years; could it (or any other significant dietary change) affect brain size on a global basis? We will come back to these questions as we investigate next how the brain works and then its metabolic and nutritional vulnerabilities.

Chapter 3

Defining Characteristics:
Vulnerability and High Energy Requirement

Embryonic Brain Development

The earliest stages of embryonic brain development involve rapid cell proliferation, resulting in accumulation of a very large number of cells destined to become neurons. The final number, form and destination of neurons depend as much on the rate and timing of programmed cell death, or *apoptosis*, as on the rate of formation of these cells. In humans, the peak number of neurons occurs at 10 to 18 weeks gestation.

Organized migration of newly formed brain cells begins soon after the proliferation phase. This migration has two important characteristics; it is time-sensitive, and different cell types migrate as separate clusters that stay in parallel alignment and become distinct layers. Cell-specific structures also start to emerge at this time, i.e. neurons grow axons and dendrites, and *neuroglia* start to differentiate into *astrocytes* and *oligodendroglia*. The latter then start to develop the myelin sheaths that surround axons.

Across animal species, there is remarkable similarity in the forms and stages through which the brain passes during its early development, i.e. the initial induction of the neural plate, followed by regionalized cell proliferation, and then cell migration and finally programmed cell death. Migration is timed such that some cells that are still developing in the deeper area of the brain where most neurons originate must then pass through zones of other neurons that have already migrated, differentiated, and are well established at their final destination. Thus, younger migrating neurons that will eventually reside in the cerebral or cerebellar cortex near the outer surface of the brain must first pass through deeper brain areas such as the thalamus, which contains older, more established neurons that will remain where they are. Apoptosis and resorption of neurons that have not migrated or adequately established dendritic connections in the layer for which they are destined is a normal phase of brain maturation.

Glia (or *neuroglia*) are the family of brain cells that support the neurons. They are mostly located between the brain's blood supply and the neurons themselves. Glia transfer nutrients and synthesize other substances necessary to maintain function of the neurons. They are also responsible for generating the myelin sheath that protects the integrity of neuronal signals and insulates each neuron's axon.

Depending on their function, the timing of neuronal maturation, i.e. development of dendrites and axons, starts earlier in gestation than for the glia and is relatively complete early postnatally. Neuronal cell division and migration is usually occurs during a ten week period at 14 to 25 weeks of gestation, after which the adult number of cells is then present in the brain. Glia start to form at about 15 weeks gestation but in contrast to neurons, they continue to divide and proliferate through adult life. Depending on the region of the brain, their maturation continues for at least two years postnatally, during which time their total number also drops dramatically.

The cerebral cortex is anatomically distinct by eight weeks gestation. By 26 weeks gestation, its multi-layered structure and the outer gray matter (mainly cell nuclei) and inner white matter (mainly myelin) are clearly evident. The different functional areas of the cerebral cortex develop at different stages according to their priority. For instance, myelination of the visual and auditory areas occurs earlier (and prenatally) than in circuits involving movement of the limbs which require integration of the cerebellum and cerebrum, and can take three to four years to complete. Control over movement of the arms and upper body develops before that of the legs, thus delaying walking over grasping with the fingers or turning the head. Final myelination of circuitry that defines more advanced decision-making and reproductive behaviour does not mature until after adolescence.

From about six months gestation until about the time of birth, the cerebrum undergoes the most rapid growth. Most of the increase in the volume of the cerebellum is postnatal and occurs more slowly than for the cerebrum. This crude measure of differential growth of brain components is consistent with the fact that the cerebellum is more concerned with control of precise movements of the limbs, an attribute which is largely lacking at birth but which becomes established over the subsequent months and years. On the other hand, it is through the mid-brain and the brain stem that virtually all incoming and outgoing signals pass, including those that control functions such as heart rhythm and lung function. Automatic control of these vital functions is needed for survival at birth so must become established fairly early in gestation.

Investing in Vulnerability

Birth is a convenient reference point at which to compare 'functional capacity' of the brain because the newborn has had no time for learning or for the influence of feeding. Within minutes of being born, some mammals can stand, use their eyes, vocalize and find the mother's nipple and feed themselves. Others, like humans, are relatively helpless and can't do much except vocalize. Therefore, based on their maturity and functionality at birth, there are two classical categories of neonates: First, are the *altricial* species, which are born helpless, immature, and usually in larger litters. Second are the *precocial* species, which are born more mature, more independent, and usually in small litters.

Rats and mice fit well into the altricial mold. Chimpanzees and other non-human primates are typical of the precocial mold. Humans fit into neither mold because they are almost always born in small litters (usually only one) yet are also born totally helpless. Pigs also break the mold because they are incredibly active and mature immediately after birth yet come in litters of eight to ten.

Clearly, what humans lack in survival capacity at birth can be made up for as they mature, but this requires 10 to 15 years to accomplish. Some can play the piano, are competitive at chess, or can solve advanced math problems at five years old. That is a spectacular improvement over what can be done at birth or what any other species can do at any time in their lives. But if the food isn't very near at hand and in a familiar place, not many five year olds can feed themselves. Few five year old humans can survive independently.

Unlike in other primates, evolution has invested heavily in what the human brain can *eventually* do in an adult but has not invested in what it can immediately do in the infant. Humans have colour and stereoscopic vision, abstract thought, language and specific physical refinements, i.e. the opposable thumb and forefinger, which greatly facilitate manual dexterity. Conversely, human evolution did not particularly invest in smell, night vision, or hearing.

The size of specific areas of the brain reflects these differences. The enlarged human cerebral cortex is used for thinking and planning, and emphasizes touch, language and vision. In contrast, there is loss of areas like the olfactory cortex, which interprets smell. Humans still use smell, probably more so than is commonly realized, but more during infancy than later in life. Memories involving smell are amongst our most enduring. Nevertheless, unlike many other animals, humans do not depend on smell to seek food or identify friend or foe.

The extraordinary capability of the human brain arises in part because of a long period of childhood that can be used to acquire language and other skills. The puzzle that needs to be accounted for in human brain evolution is that children are allowed to take so long to acquire the skills to become independent of their parents. What environmental conditions allowed humans the luxury of at least five years of total vulnerability of their infants, during which they need a guardian virtually day and night? Healthy babies don't lose their baby fat; they maintain it for two to three years so the maternal investment in putting that fat in place before birth continues long afterwards.

The very long-term investment in the human infant brain juxtaposed against its prolonged vulnerability is an unresolved issue that has to be addressed in coming to a full understanding of human brain evolution. It is a paradox in which, effectively, evolution selected for vulnerability at birth. Logically, vulnerability is the last attribute one would select for if one were constantly worried about survival.

Critical Periods – A Key Source of Vulnerability

A key attribute of normal development is that the success of each stage depends to some extent on the success of preceding events. This sequential maturation creates a series of *critical periods* in which each stage has a certain amount of time to be completed. Critical periods characterize neurological development, but also occur in development of the cardiovascular, skeletal and immune systems.

Critical periods in brain development have long been recognized both by behaviourists studying imprinting, and by neurobiologists studying the sequence of maturation of cells and processes in the brain and central nervous system as a whole. Successful completion of subsequent stages depends on (is vulnerable to) the successful completion of each earlier stage. Environmental deprivation or stimulation during the critical period for a certain event affects that specific event largely to the exclusion of other developmental processes that have already occurred.

Car assembly plants are perhaps a suitable analogy: if putting on the doors occurs before fitting the windows but the team putting in the doors doesn't get all four on before the car has moved on to the window area, then the car moves on anyway and one or more doors are left missing. In reality, the missing door would get put on later even if it had to wait until the car was completed since, because it is on the outside, a door could be fitted at any time. On the other hand, if placement of the steering wheel normally occurs before fitting the dashboard, the assembly line would have to stop if the steering wheel had not been placed in time, because it could not be done later.

If, as with brain maturation, the assembly line cannot be stopped, the car produced without a steering wheel would be useless; everything else would look right and the engine would work but it couldn't be driven. Hence, during car assembly, there is a critical time point for putting in the steering wheel, after which malfunction is assured unless the timing of this event is respected. Nowhere in infant development is the concept of critical periods in development more important than for the brain.

The difference between car assembly and brain assembly is that car assembly can be stopped. The plant manager will grumble about the cost but the only thing that is really affected is time and money. In contrast, only death can stop infant brain assembly. Death might well occur if a component of the brain as important as the steering wheel could not be installed during its critical period. The risk the brain faces is not so much one of missing parts but, still with the car assembly analogy, more like poor working conditions – bad lighting, electrical black-outs, sloppy work, or ill-fitting parts. The whole car can usually be put together regardless of the working conditions, but the number of lemons produced will be increased when assembly conditions are poor. Human brain assembly during fetal and neonatal development faces the same vulnerability.

Normal development involves the emergence and maturation of specific processes and pathways. The sensitivity of these processes to external modification has a finite critical period before or after which the opportunity for such modification rapidly declines. Developing certain behaviours or achieving a certain level of performance involves critical periods, each of which necessarily involves its own period of vulnerability. The vulnerability imposed by a critical period can be genetic or environmental and is different for different aspects of neurodevelopment. The timing, nature and severity of the insult are also important in determining its impact.

John Dobbing of Manchester University is particularly associated with the concept that critical periods in mammalian brain development arise according to the stage of overall development. After the first trimester of pregnancy, the developing brain becomes resistant to toxic agents causing morphological malformations but it remains vulnerable to Dobbing's *deficits and distortions* of function, which can remain into adulthood.

Deficits and distortions occur both within specific regions of the brain and also in the brain as a whole compared to the rest of the body. These impairments arise because of inability to complete in the prescribed order or sequence the maturation of a certain cell type or region of the brain. This may be because of defective production or migration of certain types of cells, or because of selective and excessive cell death.

Distortions arise because a deficit in a preceding stage in development affects the completion of the next stage. A distortion could involve almost any part of brain development, including a reduction in size of the cerebellum relative to the rest of the brain, a reduction in myelination caused by insufficient cholesterol required for myelin, or impaired development of reflexes or a certain behavioural pattern. Distortion in brain myelination (*hypomyelination*) is one of the main consequences of postnatal malnutrition that may well account for the significant decrease in brain size and function caused by malnutrition.

Vulnerability and Resistance to Brain Damage

The risk associated with interrupting the brain's oxygen and energy supply is permanent damage because, unlike some organs such as muscle, skin or adipose tissue, the brain is committed to a continuous state of high activity; it never shuts down. Some areas of the brain become more active with stimulation but the brain as a whole is still considered to be a continuously active organ. Interruption of the oxygen supply to the brain prevents renewal of brain fuel needed for the ion gradients and neurotransmitter release that are the basis for signal transmission. Fuel stores within the brain last only a few minutes so inability to replace them means imminent cell death and permanent damage to the affected region.

The vulnerability of the brain to permanent damage is common to all vertebrates but some species have extreme resistance to such damage. Goldfish can withstand oxygen-free (*anoxic*) water for sixteen hours at 20°C without noticeable loss of brain function. Some carp can withstand anoxic water at near freezing temperatures for months. Some freshwater turtles can withstand anoxia for at least three months at 3°C.

Unlike humans, goldfish, carp and turtles are not characterized by particularly advanced cognitive capacity. Humans and goldfish are therefore examples of two extreme strategies in brain evolution (Table 3.1). In goldfish, there was minimal evolutionary investment in cognition and maximal investment in protecting the brain against various environmental insults such as low oxygen, low temperature or poor nutrient availability.

In humans, the investment was the opposite: the brain's adaptability or plasticity got maximum effort but the trade-off was that it remained highly vulnerable to insult and defective development. If protection of the brain is the priority, it is because environmental factors - oxygen, temperature or nutrition - are variable and often insufficient. The cost of that priority is that brain size must remain small and control only absolutely essential functions so that nutrient and oxygen demands are manageable even under difficult circumstances.

In contrast, if vulnerability in brain development has arisen it is because, throughout millions of years of evolution, those parameters on which 'optional' brain functions depend, i.e. advanced cognition, have been reliably available and vulnerability to them is only rarely exposed. Lack of oxygen, hypothermia or nutritional insult still cause serious problems in the species with the vulnerable brain but the chances of exposure to such circumstances are markedly lower than in the goldfish, carp or turtle.

Table 3.1. Resistance or plasticity: the two extreme strategies of brain evolution.

	Goldfish	Non-Human Primate	Human
Variability of environment:	High	Low	Low
Resistance to damage:	High	Medium	Low
Vulnerability of the brain:	Low	Medium	High
Investment in cognition:	Low	Medium	High
Plasticity of brain function:	Low	Medium	High
Need for stable fuel supply:	Low	Medium	High
Investment in fuel reserves:	Low	Low	High
Functionality at birth:	High	High	Low

Mammals, especially primates, have had dependable oxygen and nutrient supply for tens of millions of years. Genes refining and extending brain development have therefore evolved without the constraint of constant vigilance against environmental insult. Hominids took it one step further than did primates and humans took it the furthest. Hence, brain vulnerability necessarily co-evolves with increasing cognitive capacity. Both are only possible with exposure to a very stable supply of oxygen, good control of brain temperature and rich nutrient supply.

Because uterine contractions tend to compress the baby and restrict blood flow in some areas of the body during birthing, transient under-oxygenation of brain *(hypoxia)* commonly occurs during birth. The lower energy requirement of the infant compared to the adult brain probably helps it better tolerate hypoxia. For instance, during recovery from hypoxia in infants, lactate enters the brain in proportion to its concentration in blood and is immediately used as a fuel. This spares glucose, accounting for the paradoxical rise in brain glucose after hypoxia. Human infants have some brain glycogen but, in contrast to adults, brain glycogen in the infant seems unresponsive to hypoxia.

Of the three major environmental insults to the brain (lack of oxygen, low temperature, or nutrient lack), only nutrition is really relevant to human versus non-human primate evolution. Protecting the brain against vulnerability means either cognitive sophistication must be restricted or better protection of the brain must be assured in the form of brain fuel reserves. By evolving body fat stores during fetal development, humans developed better nutrient and fuel protection for the brain than did non-human primates or any other terrestrial animals.

Human infants are much less functional than are the infants of many other precocial species including non-human primates: at birth, they cannot sit, stand, walk, grasp, etc. By permitting the lowest possible 'functionality' at birth, this put the lowest possible energy demand on the fetal and neonatal brain, thereby reducing oxygen consumption and hence vulnerability. By developing the lowest possible functional demands on their infants while also obtaining best quality diet available, hominids limited the vulnerability of their brains more than anything else could (Table 3.1).

As a result of a lengthy period of evolution in a highly stable environment, the human infant as a whole became increasingly vulnerable. Low functionality of the infant places fewer demands on the brain but makes the infant increasingly dependent on the mother. Natural selection considers not what the brain or other organs would like but what the organism as a whole can tolerate. Infant viability must not have been substantially imperiled or humans would not have evolved cognitive advancement. An environment providing a rich and stable nutrient supply would have offered the best opportunity to combine brain expansion and extreme infant vulnerability.

It would have been advantageous if the more advanced human brain could have evolved a lower fuel requirement. That also would have made it less vulnerable, but this would have been at the cost of limiting the expensive process of establishing and maintaining brain connectivity, which is the hallmark of advanced cognitive capacity. In effect, without having evolved adequate fuel reserves for the brain, non-human primates were restricted to a lower functionality, lower vulnerability strategy of brain development.

How Much Energy is Needed for Brain Activity?

It is difficult to measure how much energy brain cells need when they are active. Is there such a thing as an inactive brain cell or brain region? If there is a difference caused by activity, how is it measured? For species with higher cognitive aptitudes, what is the energy cost of that increased ability? Underlying this problem is a general assumption that active brain cells use more energy than inactive brain cells.

Brain imaging tools like functional magnetic resonance do not measure neuronal activity *per se* but, rather, they image blood flow to brain regions. Positron emission tomography is another common technique to measure brain energy metabolism and it, too, does this indirectly by measuring regional brain uptake of glucose. Both these techniques show that increased blood flow and increased glucose utilization accompany an increase in activity of a certain area of the brain, i.e. the visual cortex.

The increased energy demand of higher brain activity is usually met by increased oxidation of blood sugar (*glucose*). Increased blood flow is a response to a combination of related events occurring during brain activity, including increased glucose utilization, lower oxygen levels, and increased carbon dioxide production. Blood flow is what can be measured and the common assumption is that changes in blood flow to a specific region of the brain are an accurate substitute for changes in brain activity.

Several years ago, Marcus Raichle of Washington University challenged this view. His imaging data confirmed that blood flow and glucose uptake increased in brain areas that were active. However he also showed that there was less than a 6% increase in oxygen consumption in the active area. Raichle suggested this small increase in oxygen consumption was insufficient to account for brain activity. In fact, in some of his experiments, oxygen extraction from the blood seemed to *decrease* during visual and somatosensory stimulation.

Therefore Raichle reasoned that if blood flow and glucose uptake increased yet oxygen uptake decreased in the affected area, oxygenation of the active brain area must paradoxically increase not decrease during activity, i.e. that brain activity does not depend on an immediate increase in oxygen uptake. Thus, it was unlikely that neural activity alone is sufficiently costly to increase both oxygenation and regional brain blood flow.

The basis for Raichle's controversial argument is that the brain is already using the maximum oxygen available just to maintain membrane potential, ion gradients, neurotransmitter production, etc. If brain oxygen uptake were already maximal in the 'resting state', added brain activity would have to be independent of oxygen uptake. Therefore, lactate (or an alternate fuel) would have to be produced to provide the extra energy substrate during stimulated brain activity. His results supported this but the rise in lactate seemed too low to really account for the increased energy requirement of the stimulated region of the brain.

Some labs showed that oxygen uptake increased more with the same test as Raichle used. Others supported Raichle's findings but showed that although tactile stimulation was not sufficient to increase oxygen uptake, visual stimulation was. Raichle's data were also supported by reports showing that brain lactate increased during brain stimulation in the cat. A rise in lactate usually suggests that the tissue is active without using oxygen. These inconsistencies support the need for a reexamination of what actually fuels brain activity.

Of the apparently low amount of oxygen consumed during brain activity, as much as half seems necessary to fuel the energy needs of glia, which are the support cells not directly implicated in neural signal transmission. An increase in blood flow may also help maintain potassium homeostasis and pH. Hence, the increase in oxygen consumption alone during brain activity is probably insufficient to account for increased blood flow in an active brain area. Raichle felt that integration of the activity of glia and capillaries with that of active neurons in a specific area of the brain may well account for the blood flow and increased oxygen consumption of that region.

Hemoglobin is the oxygen transport molecule in the blood. With activation of a brain area there is a rapid loss of oxygen from hemoglobin causing a transient rise in *deoxyhemoglobin* (hemoglobin having no oxygen) lasting 200-300 milliseconds. About 3 seconds after stimulation, the situation reverses with an increase in *oxyhemoglobin*. This confirms that the brain rapidly consumes oxygen, which is then replaced with more oxygen from the blood. The problem is that the techniques used in these experiments do not show which cells are using the oxygen, or glucose for that matter.

Overall, these studies implicate other brain cells besides the neuron itself in neural communication, a point that Raichle was initially at pains to make. Some of the glucose taken up by the brain is probably made into lactate by *astroglia*, which the neurons then use to make ATP during a burst of activity. Lactate transporters and other enzymes exist making this more indirect scenario feasible.

Energy Requirements of the Brain

KW Cross and D Stratton at the London Hospital Medical College showed that the temperature deep in the ear indirectly measures the brain's heat production and is a good estimate of the brain's metabolic rate. These and other studies were supported and extended by Malcolm Holliday at the University of California at San Francisco, who showed that the newborn infant brain takes about 74% of the energy consumed by the whole body (Table 3.2). By the time an infant weighs about five kilograms, i.e. at about three months old, the percentage of whole body energy intake consumed by its brain has dropped to about 64% of the total. This compares to the liver taking 18%, muscle taking 7%, and the other organs including the heart competing for the remaining 11% of energy intake.

Table 3.2. Energy requirements of the human brain from birth to adulthood (modified from Holliday, 1971).

Age (years)	Body Weight (kg)	Brain Weight (g)	Brain's Energy (kcal/day)	Consumption (% of total)
Newborn	3.5	400	118	74
4-6 months	5.5	650	192	64
1-2 years	11	1045	311	53
5-6 years	19	1235	367	44
10-11 years	31	1350	400	34
14-15 years	50	1360	403	27
Adult	70	1400	414	23

If the baby is born prematurely or was malnourished *in utero*, muscle and fat are smaller but brain weight is close to normal, so an even higher percent of total energy consumed by the body goes to the brain. There has to be an upper limit to the amount of energy that can be drawn by the brain, even in the most premature the infant. For example, in a premature infant weighing 1100 grams (the average being 3500 grams) with a brain weighing 190 grams, the same metabolic demand by the brain as in a term infant would exceed the energy intake of the whole body. Therefore, the brain's metabolic rate must either ramp up postnatally or be significantly compromised by premature birth. Both may be true but exact values are not known.

In adults, the body's energy requirement is about 25 kilocalories per kilogram of body weight, or 1700-1800 kilocalories per day. Holliday showed that the energy requirement of the brain, heart, kidneys and liver is completely disproportionate to their size; they consume two thirds of the body's energy intake but they represent less than 6% of the body's total weight. In terms of energy consumption (kilocalories per kilogram of tissue per day), the metabolic rate of the heart is highest at 610. The heart is followed by the kidneys at 390, the brain at 296, and the liver at 290 kilocalories per kilogram per day. Well below these values are those of other organs, including skeletal muscle at about 17 kilocalories per kilogram of tissue per day.

Because of its relatively large size compared to the heart and kidneys, the adult human brain consumes the most disproportionate amount (about 23%) of the body's total energy needs. The heart and kidneys are metabolically more active but the brain takes a higher proportion of total energy consumption because it is larger than the heart and kidneys. Thus, if they were the same size, the brain would have a significantly lower energy requirement than the heart or the kidneys. Skeletal muscle has a relative low rate of energy consumption but, overall, still takes 25-30% of the body's energy intake because it occupies 40-50% of total body weight and represents 80% of the body's total cell mass.

Experiments have been conducted to measure the energy cost of the brain's metabolic activity. These experiments either temporarily knock out the sodium-potassium pump with a drug or they hyperstimulate the brain using seizures. They show that the combined demand by the sodium-potassium pump and synaptic neurotransmission account for 50-60% of the brain's energy consumption. The remaining energy requirement of the brain is for synthesis of neurotransmitters and numerous structural and regulatory molecules.

Still, about 20% more glucose is consumed by the brain than can be strictly accounted for by the brain's energy needs. This excess breakdown of glucose by the brain appears in newly synthesized lactate and pyruvate, which can be measured leaving the brain in venous blood. Some glucose carbon also goes into other products in the brain including neurotransmitters.

Unlike in the heart, muscle or liver, there are clear differences in oxygen consumption across brain areas. Glucose consumption is higher in the brain areas with the highest metabolic activity, which are the *auditory cortex* and *inferior colliculus*, both of which are used for hearing (see Table 3.2). It is interesting to note here the relationship between brain energy metabolism, iodine and hearing: The development of hearing is the most energetically expensive area of the brain and iodine is a key nutrient in the regulation both of energy metabolism and the normal development of hearing (see Chapter 6). Expansion of the auditory region in humans undoubtedly facilitated the evolution of speech uniquely in humans, so two key aspects of human evolution (speech and brain enlargement) have marked energy demands and are vulnerable to iodine status.

Table 3.2. Oxygen consumption (micromoles/100 grams/minute) in different areas of the monkey brain.

GRAY MATTER:		
	Inferior colliculus	103
	Auditory cortex	79
	Superior olivary nucleus	63
	Medial geniculate	65
	Vestibular nucleus	66
	Mamillary body	57
	Thalamus (lateral nucleus)	54
	Cochlear nucleus	51
	Parietal cortex	48
	Caudate-putamen	52
	Thalamus ventral nucleus	43
	Visual cortex	59
	Cerebellar nucleus	45
	Lateral geniculate	39
	Superior colliculus	55
	Nucleus Accumbens	36
	Hippocampus	39
	Pontine gray matter	28
	Globus pallidus	26
	Substantia nigra	29
	Cerebellar cortex	31
	Hypothalamus	25
	Amygdala	25
WHITE MATTER:		
	Corpus callosum	11
	Internal capsule	13
	Cerebellar white matter	12
WEIGHTED AVERAGE		36

Data compiled from Sokoloff (1991).

One of the earliest approaches to determining the energy requirements of the brain was by comparing the concentration of glucose, oxygen and carbon dioxide in the *carotid arteries* supplying blood to the brain with their concentration in the *carotid veins* coming from the brain. The difference between the arterial and venous concentration of a substance is the *arterio-venous difference*. If the arterio-venous difference is a positive number, the tissue is taking up the substance in question because more arrives at the tissue in the arterial blood than leaves in the venous blood, i.e. oxygen or glucose. If the arterio-venous difference is negative, the tissue is producing more of that substance than arrives in the arterial blood, i.e. carbon dioxide.

The high volume of blood flowing through the brain means that useful measurements of arterio-venous difference in the brain are restricted to metabolites in high enough concentration, mainly oxygen, carbon dioxide and glucose. Numerous arterio-venous difference measurements for glucose and oxygen show that for each molecule of glucose consumed by the brain, about 5.5 molecules of oxygen are also consumed.

Oxygen consumption by the brain can vary with changes in blood flow, i.e. with dilatation of cerebral blood vessels in response to low oxygen or high carbon dioxide load. Extraction efficiency of oxygen can also change irrespective of a lack of change in blood flow. Still, arterio-venous differences are sufficient to calculate the brain's energy (glucose) consumption, which is about 3.5 grams per 100 grams brain per minute in adults. Glucose consumption is even higher in the infant than it is in the adult brain (5.3 grams per 100 grams brain per minute).

Brain energy requirements appear not to change with changes in 'normal' brain activities as different as sleep or activities requiring intense concentration. However, the brain's energy requirement is measurably reduced during loss of consciousness. It may also decrease in the elderly and in conditions in which the blood's oxygen carrying capacity is reduced, such as anemia. There is disagreement on whether hypothyroidism reduces brain oxygen consumption. At levels raising blood pressure and causing anxiety, adrenaline notably increases oxygen consumption by the brain. Convulsions and epileptic seizures can double the brain's energy consumption. Hyperthyroidism (overactive thyroid gland) has no measurable effect on the brain's consumption of energy, even though it can raises energy expenditure of the body as a whole by up to 50%.

The Brain's Energy Substrates Linked to Glucose

With normal food availability, the brain runs almost exclusively on glucose. Glucose metabolism is normally *aerobic* (oxygen consuming). *Lactate* is a three-carbon molecule derived from glucose metabolism by a process called *anaerobic glycolysis* or glucose breakdown without using oxygen. In most tissues, lactate can function as a temporary alternative fuel when insufficient oxygen is available for normal glucose metabolism, i.e. during hypoxia. Hypoxia occurs routinely during extreme exercise, especially if one is insufficiently trained.

Because of uterine contractions and compression of the blood supply to the baby, transient brain hypoxia also commonly occurs during birth. The lower energy requirement of the infant brain compared to the adult brain helps it better tolerate hypoxia. This is supplemented by the brain's ability to use lactate as a fuel, which enters brain from the blood in proportion to arterial concentration.

During prolonged hypoglycemia, i.e. starvation, when there is no incoming glucose or other carbohydrates, the body is capable of using *gluconeogenesis* to make glucose from several other substances (see Figure 3.1). In muscle, removal of the nitrogen from amino acids makes these molecules good short-term sources of carbon for glucose. This use of amino acids to make glucose consumes muscle protein and leads to muscle breakdown in starving humans so, at best, it is a temporary solution.

Figure 3.1. Compounds contributing to gluconeogenesis.

Glucose can also be produced from recycled lactate or pyruvate. This source doesn't represent net glucose synthesis because the lactate or pyruvate came from the breakdown of glucose in the first place. Glucose can also be made from *glycerol* produced during the breakdown of triglycerides from body fat. Glycerol has three carbons and readily slips into the pathway of glucose synthesis. Gluconeogenesis from these other molecules works for 24-48 hours but is an inadequate solution to meeting the constant and high energy needs of the brain. Amino acids come principally from muscle so their breakdown to supply the brain's energy needs for even a few days would be incapacitating. Thus to provide longer-term back-up fuel requirements for the brain, a different energy substrate not involving glucose or tissue breakdown is needed.

Ketone Bodies – Alternatives to Glucose

Ketone bodies (*ketones* for short) are that different energy substrate and they are derived from fatty acid oxidation (see Figure 3.2). There are three ketones – *beta-hydroxybutyrate, acetoacetate,* and *acetone.* All the steps in ketone metabolism occur in the brain, including conversion of beta-hydroxybutyrate to acetoacetate, acetoacetate to *acetoacetyl CoA,* and acetoacetyl CoA to *acetyl CoA.* Acetyl CoA is the final common denominator in the metabolism of all the different energy substrates for ATP production. Conversion of beta-hydroxybutyrate to acetoacetate occurs via the enzyme – *beta-hydroxybutytrate dehydrogenase* - which is very active in both mitochondria and synapses of nerve endings. It is also present in glia, neurons and nerve fibers.

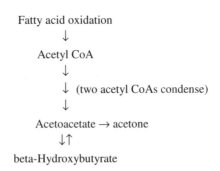

Figure 3.2. Ketone body synthesis. Multiple arrows in a row indicate several steps; two-way arrows indicate reversibility of the reaction. Acetoacetate and beta-hydroxybutyrate both have four carbons. Acetone has three carbons and is formed by the spontaneous decarboxylation of acetoacetate.

Brain uptake of ketones increases in proportion to the ketone levels in the blood. As fasting starts, brain ketone uptake is modest but it increases the longer that fasting continues, i.e. the longer that elevated blood ketone levels are sustained. When fasting starts, the initial rate of brain ketone body uptake is relatively slow because plasma ketone concentrations are low, and because the *monocarboxylic acid transporter* is still not fully activated for ketone transport. As fasting continues, this all changes because the stimulation of insulin production by dietary carbohydrate stops and so production rises and brain uptake of ketones becomes fully activated. When insufficient glucose is available, ketones can supply up to two thirds of the adult human brain's energy needs.

Ketone utilization by the brain is not an adaptive response because the amounts and activities of the relevant enzymes are not changed by starvation and always exceed the amount necessary to supply the brain's energy needs. This implies that the brain is used to metabolizing ketone bodies and is prepared to use them as soon as they are available. The factor that limits ketone body use by the brain is therefore ketone availability and not the brain's ability to use them. Ketone availability is determined by glucose and insulin levels and the ability to generate ketones, i.e. fat stores.

Ketones seamlessly replace an inadequate supply of glucose to meet the brain's energy needs, but some glucose is still needed to supply at least one third of the brain's energy needs. In an experimental model involving marked ketosis, the rat brain shows no electrical activity when the blood glucose concentration is below a certain point (less than two millimolar), so high levels of ketones are unable to completely substitute for glucose in the brain.

Thus, during starvation when insulin and glucose are low and blood ketones are elevated, the brain still has an absolute dependence on glucose for about one third of its energy requirement. Hence, the brain actively transports glucose as well as ketones even during starvation and fully developed ketosis. In addition to providing the carbon to replace oxaloacetate, glucose is essential for the brain as the precursor to lactate, which is exchanged when ketones are taken up by the brain.

There are two key advantages to having ketone bodies as the main alternative fuel for the human brain. First, humans normally have significant body fat stores so the supply of fatty acids to make the ketones is almost unlimited. Second, using ketones for up to half the brain's energy requirement frees up a lot of glucose and greatly reduces the risk of detrimental muscle breakdown or compromised function of other organs dependent on glucose. Ketone body uptake by the brain is five times faster in newborns than adults, and four fold faster in older infants compared to adults, so the infant brain is arguably better equipped with 'fuel insurance' than the adult brain.

Radiotracer studies show that, compared to glucose, ketones are unevenly utilized by the brain. Beta-hydroxybutyrate penetrates all of the cerebral cortex but more beta-hydroxybutyrate than glucose penetrates the sub-cortical areas. Low amounts of labeled beta-hydroxybutyrate are found in certain deep regions of the brain, but are elevated in the hypothalamus. By far the most labeled beta-hydroxybutyrate is found in the *pituitary* and *pineal glands*, both of which have endocrine regulatory functions. The region-specific brain uptake of ketones suggests they may be able to supply certain regions but might not be capable of supplying the energy requirement of the whole brain.

The ability of the brain to switch to ketones as the primary energy substrate is not unique to humans but is better in humans than in other omnivores like the rat, pig, or monkey. Adult carnivores like dogs achieve negligible ketogenesis during fasting but also have smaller brains and markedly lower brain energy needs.

Challenges to Meeting the Brain's Metabolic Requirements

Several challenges for brain energy metabolism arose during its evolution (Table 3.3): The first challenge is the brain's high metabolic rate. As in other cells, to regenerate ATP, the brain has to process glucose through the *tricarboxylic acid cycle*. *α-Ketoglutarate* and *oxaloacetate* are two key intermediates in the cycle that are normally in equilibrium with the amino acids - *glutamate* and *aspartate*. Glutamate is also in equilibrium with other amino acids and other related metabolites capable of being converted to glutamate. This interconvertability of metabolites is useful, indeed necessary, if, like the brain, the tissue has high energy needs and an active tricarboxylic acid cycle. The brain uses the tricarboxylic acid cycle to generate its ATP but the problem is that same glutamate entering the tricarboxylic acid cycle is also a neurotransmitter that can trigger undesirable signals if its concentration in the brain becomes raised inappropriately.

Table 3.3. Unique challenges to brain energy metabolism.

1.	The maintenance of the membrane potential and synaptic function of neurons necessitates high energy consumption.
2.	The blood-brain barrier markedly limits access of potentially suitable energy substrates such as fatty acids. Despite its ability to oxidize fatty acids, the brain's need for a strictly controlled fatty acid composition makes these otherwise excellent energy substrates largely inaccessible to the brain.
3.	The brain uses glucose as its principle energy substrate but is insensitive to insulin, the principle hormone controlling glucose metabolism in other tissues.
4.	Glycogen can only be a minor fuel reserve for the brain because there is insufficient space within the cranium for the necessary tissue expansion required by the additional water that accompanies glycogen production.

The brain is therefore forced to maintain about a thousand-fold lower glutamate concentration in the fluid surrounding the cells where signals get generated and transferred among neurons than inside its cells where the brain can use glutamate to generate ATP without risking an undesirable electrical discharge.

The second challenge is the blood brain barrier. In terms of nutrient and metabolite in-flow, the brain is almost sealed off from the rest of the body. Still, it uses metabolites found elsewhere in the body, including certain amino acids (GABA, glycine, and glutamate) that are both protein constituents and neurotransmitters. Liberal access by amino acids that are neurotransmitters but coming from outside the brain would destroy the tight, local control of their release and would severely disorganize brain function so their access to the brain has to be extremely carefully regulated.

Complex molecules, including albumin and lipoproteins, can dock at receptors on the capillaries of the brain but the blood brain barrier blocks direct access to the brain by these large molecules. Specific transporters have evolved for the brain to take up the exogenous substances it needs, even for simple but important molecules like glucose.

The third challenge is the brain's use of glucose as its main energy substrate. Unlike other tissues, brain glucose uptake is not responsive to high or low insulin in the blood, nor does brain glucose uptake increase when blood glucose is elevated. This relative insensitivity to insulin helps maintain a steady rate of brain energy metabolism by preventing wide swings in brain glucose uptake. The brain can somewhat increase the efficiency of glucose uptake when blood glucose is low, but this process seems to become less efficient in elderly humans and may contribute to decaying cognitive function and memory loss.

Fourth, it was encased in cartilage or bone, thereby precluding significant daily fluctuations in size. Unlike in muscle and liver, this prevented significant storage of glucose to use as an energy reserve. *Glycogen* is the storage form of glucose but glycogen synthesis requires three to four grams of water for each gram of glycogen, which takes a lot of space. This is fine for liver and muscle, which are not encased by bone but cannot be accommodated by the brain. In fact, there is a small amount of glycogen in the brain but it appears to last for no more than a few minutes and so takes up very little space. In effect, sufficient glucose to meet the brain's needs cannot be stored by the brain so it is dependent on glucose supplied to it in the blood.

A few minutes supply of glycogen is present in the brain but this small reserve is of no long-term use. If the brain depended only on glucose, its own energy needs when the body was low in energy reserves, i.e. starved, would consume significant amounts of amino acids that can be converted to glucose and would destroy the protein in other organs like muscle and intestine. Hence, a final challenge is that the brain needs alternative energy substrates to glucose that are compatible with the body's energy needs as a whole.

Energy Needs and Human Brain Evolution

Taking account of the need to constantly supply large amounts of energy to the brain is fundamental to developing any reasonable explanation of human brain evolution. This constant need by the brain for fuel exists regardless of conscious (adult) brain activity or immaturity of the brain, i.e. in the infant. The human neonate is a clear example of natural selection investing a lot of effort in an organ that provides very little short term benefit; human babies do very little cognitively and yet this genetic and metabolic investment has been heavily made in humans.

It is necessary to explain how this added dietary energy started to became available to early hominids but not to other species, especially non-human primates. Such an explanation has to include a mechanism by which abundant dietary energy supply became consistently available. In addition, it is necessary to explain how the extra dietary energy was available not only to meet the needs of the brain but also to deposit body fat in the fetal and neonatal period.

Chapter 4

Fatness in Human Babies:
Critical Insurance for Brain Development

When hominids first evolved body fat is unknown. Without clues in the fossil record or other artifacts no one can say with any certainty what degree of fatness was present during the long period before enduring visual images of humans were first made tens of thousands of years ago. What is clear however from the earliest sculpted or carved images seen as representing the human (mostly female) form is that at the dawn of modern humanity 20,000 to 30,000 years ago, women were fat.

Many of the sculpted or etched images show fatness equivalent to that of a mature, obese woman. Others, such as the etching recently discovered at a cave in Cussac near the Dordogne River in southern France, clearly show a firmer, less layered fatness more typical of a younger woman. Perhaps fat women were uncommon or some artistic license was taken. However, bears, horses, mammoths, reindeer (and salmon; see cover) drawn on or etched into cave walls were neither fat nor badly proportioned, so some awareness of fatness was part of and apparently peculiar to the human experience at that time.

Assuming that many young and older women truly were fat at the time behaviourally modern humans first evolved, it seems reasonable that infants would have been similarly fat. In any event, fatness is characteristic of healthy infants today (Figure 4.1). It is a reasonable assumption that fatness in human babies is at least 20,000 years old, but probably much older. In all likelihood, body fatness was part and parcel of becoming human and, like large human brains and bipedal

Survival of the Fattest

locomotion, probably took a long time to emerge. The problem is that it is impossible to quantify the degree of fatness of pre-humans because no known attributes of fatness are preserved after death.

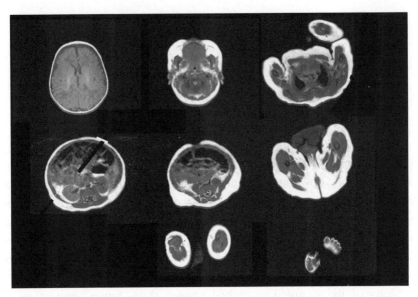

Figure 4.1. Body fat distribution on a healthy human newborn infant as shown by cross-sectional scans obtained using magnetic resonance imaging (modified from Harrington *et al* 2002, with permission).

Body fat is widely claimed to have evolved in humans either as a form of insulation against the cold or as a physical cushion for more comfortable sitting down. Body fat may do these things but the primary reason it exists in any organism is to be a storage form of excess incoming energy from the diet. Body fat stores expand when more dietary energy is consumed than is needed to stay alive and shrink when energy intake is less than energy expenditure. Fat must have evolved in humans for fundamentally the same reason – at some stage, hominids must have had access to a modest excess of dietary energy relative to energy expenditure.

In affluent countries, there is great concern today about the health risks of excess fatness in adolescents and adults. This concern has led to extensive research into how fatness develops in adolescents and adults and why so many people have lost the appropriate balance between appetite, food selection and physical activity. As a consequence, obesity has increased in wealthier countries to what many now call epidemic proportions. Obesity brings along with it a cluster of heightened health risks including premature birth, hypertension, insulin-resistant diabetes, premature cardiovascular disease and cancer.

In light of the very legitimate concern about the health risks and insidious rise of obesity in developed countries worldwide, the fact that fatness is a vital form of energy insurance for healthy, normal infant development is commonly forgotten. Contrary to the adult situation where excess fatness is a harbinger of a health risk in later life, fetal and neonatal fat is beneficial, indeed essential, for normal growth. Fatness in newborns is distinct from excess fat in adolescents and adults.

Energy Balance

What is referred to as *dietary energy* comprises the energy that can be obtained from food and used to make energy (ATP) for cells to function. For their weight dietary fats have twice the energy content of dietary proteins, sugars or starches. Food is converted to energy for essential processes like breathing, heart contraction, immune surveillance, and organ function in general. This minimal necessary energy requirement for life is called the *basal metabolic rate*, which allows an organism to meet the barest essentials of life, i.e. while asleep or totally inactive. It does not include the energy requirements of physical activity.

All daily activities require additional energy than the amount used during sleep because activity requires muscles to work. Oxygen and fuel consumption increase to power the muscles, so heart and respiratory rates are also higher during activity. The digestive system converts dietary constituents into useable *energy substrates* but the processing of food to meet the added fuel requirements also contributes to energy expenditure.

These metabolic processes consume more energy than is needed while asleep so additional dietary energy (food) has to be consumed in proportion to the amount of additional activity. Depending on activity level, average total daily energy expenditure exceeds basal metabolic rate by 50-100%. This means that someone who is extremely active may need double the energy intake of an inactive person. Similarly, if one is pregnant, nursing, or fighting a severe infection, more dietary energy will be required for these additional activities; if it isn't consumed, this additional energy will have to come from body stores. Many people in affluent Western societies eat as though they are quite active but actually lead sedentary lives, so body fat gradually accumulates because energy intake from the diet continually exceeds the body's needs.

Glucose meets most short-term energy needs. Because of its primary importance as a fuel, in addition to liberating it from sugars or starches in the diet, glucose can also be made in the body. Glucose is a 'simple sugar' or *monosaccharide*. Table sugar or *sucrose* is a 'double sugar' or *disaccharide* of glucose and another simple sugar. Starches are *polysaccharides* or chains of many sugars, some of which are glucose. After digestion and absorption, incoming glucose has several fates. It may go straight into the tricarboxylic acid cycle where it will be used to make ATP. It may also be bulk stored in liver or muscle as *glycogen*, which is a molecule comprising chains of glucose. If glycogen stores are full, remaining glucose could also be converted into amino acids or fatty acids. The important point is to get excess glucose out of the blood, a process that is orchestrated by the hormone, *insulin*.

When less energy is consumed than the body needs, hunger is experienced and one eats to restore the *energy balance* between incoming and outgoing fuels. If one has not eaten within the past 24 hours, there will have been no incoming glucose in any form. However, the body's need for fuel will not have changed so glycogen releases glucose stores to continue to meet those fuel needs. Depending on activity level, glycogen stores may last 24 hours. As glycogen reserves dwindle, the continuing low insulin levels lead to recruitment of other fuel reserves other than glycogen.

During *gluconeogenesis,* proteins in muscle provide amino acids to convert to glucose. However, fatty acids from body fat are the main alternative fuel to glucose. The various body fuel reserves can potentially keep a healthy adult human alive for up to 30-40 days of total starvation. Thus, there is a constant interplay between glucose supply and demand, with short- and long-term alternate fuels filling in when glucose stores are used up.

Body Fat Distribution

Fat is present in two general regions of the body: *Subcutaneous fat* is under the skin while *visceral* or *intra-abdominal fat* surrounds the abdominal organs. There are many other smaller fat stores such as those around the kidneys or heart, or in between the large muscles of the arms and legs, but the main places are the subcutaneous and visceral sites. By definition, visceral fat is restricted to the abdominal cavity itself. Depending on the degree of fatness, subcutaneous fat is prominent around the abdomen but is also found on the buttocks, chest, shoulders, back, upper arms and legs, cheeks, and under the chin.

Two general body profiles of fat accumulation are widely recognized in obese adult humans. The *apples* accumulate fat mostly around the waist but less on the buttocks and thighs. Men with beer bellies are classic examples. The *pears* have the opposite pattern with fat accumulating more on the buttocks and thighs and less on the abdomen. The apple pattern of excess fatness is more of a health concern because it is associated with more visceral fat and a higher risk of *insulin resistance* or *non-insulin-dependent diabetes* (also called Type 2 or adult-onset diabetes).

Non-insulin-dependent diabetes is of great concern as a cardiovascular disease risk especially since it is no longer restricted to adults but is now seen in obese adolescents and even children. The number of people with non-insulin-dependent diabetes is rising steadily in virtually all areas, city and countryside alike, where people eat in excess of their energy needs and are insufficiently active.

By looking at the body from the outside, it is not always apparent what proportion of the fat on the abdomen is due to visceral or subcutaneous fat. Sophisticated medical imaging techniques now make it possible to measure the thickness and location of fat by images of the body taken in cross-section. Those who successfully lose weight still may not be fully satisfied with their body shape but these images almost always show that they are healthier if they are losing weight from the visceral site, i.e. that their risk of insulin resistance and heart attack has gone down.

Uniqueness of Human Baby Fat

Perhaps because excess fat is considered to be unattractive in adults, its physiological and evolutionary significance in babies has largely been missed. We are attracted to plump babies because they appear healthy and their fatness somehow makes them innately attractive. Nevertheless, few people have considered the biological origins and possible evolutionary value of fatness in babies. The fossil record has yielded few specimens of hominid infants. In addition, body fat doesn't fossilize. This makes almost everything about the emergence of fatness in hominid infants completely speculative.

In 1993, I published a paper co-authored with Michael Crawford and Laurence Harbige in which we speculated that since body fat was a key form of energy insurance for the voracious energy needs of the developing human brain, baby fat might have had a key role in human brain evolution. Not long afterwards, in *The Descent of the Child,* Elaine Morgan noted that the ten fold higher number of fat cells in humans than other land animals quite reasonably needs explanation. She also recognized the uniqueness of human baby fat and speculated that it would have both advantages and disadvantages.

One advantage of fat would be that if some hominids had made liberal use of shore-based foods, they would have accessed the water and more body fat would have facilitated flotation. Morgan, too, recognized that it would be useful to have a fat store for the costly energy requirement of the developing brain. However, larger fat stores have the disadvantage of making childbirth more difficult because human babies are bulkier than are other primate babies. In addition, it is metabolically costly to deposit fat in neonates.

Human baby fat is different from adult body fat in at least four important respects (Table 4.1): First, its distribution is different. Baby fat is much more evenly distributed on the limbs, face and the torso. Adults tend to accumulate fat centrally and have little fat on the lower arms or legs, or on the hands or feet, although distribution at these sites increases with extreme obesity. Second, magnetic resonance imaging studies show that baby fat is almost all subcutaneous and, unlike in adults, little or none is present around the visceral organs.

Third, in comparison to adults, baby fat at birth contains very low amounts of the two primary dietary polyunsaturated fatty acids, linoleic acid and alpha-linolenic acid. These two fatty acids are typically found in adult body fat in amounts corresponding to the amount consumed. However, baby fat at birth has almost none of these two vitamin-like fatty acids so the fetus accumulates very much less linoleic acid and alpha-linolenic acid compared to the significant place they occupy in the mother's fat. Hence, before birth, the placenta somehow screens out these two fatty acids and sends very low amounts of them to the fetal body fat that is developing in the third trimester.

Table 4.1. Important differences between infant and adult body fat.

1. Baby fat has wider and more even distribution under the surface of the skin.
2. Baby fat is almost exclusively under the skin and, unlike in adults, is not present in the abdominal cavity as *visceral* or *intra-abdominal fat*.
3. Baby fat contains *very low* proportions of the two most common dietary polyunsaturated fatty acids – linoleic and alpha-linolenic acids.
4. Baby fat contains *significantly higher* proportions of the two key long chain fatty acids in membranes – arachidonic acid and docosahexaenoic acid – than in adult body fat.

Interestingly, the brain's fatty acid composition is like that of the baby's body fat in also having very low amounts of linoleic acid and alpha-linolenic acid. This is unlike adult body fat or all other organs whether in babies or adults, which tend to have 10-20% of their total fatty acids as linoleic acid. Thus, it seems undesirable, uneconomical, or both to accumulate more than trace amounts of linoleic and alpha-linolenic acids in body fat before being born.

The fourth reason baby fat is different from adult fat is that it contains higher amounts of the two principal *long chain polyunsaturated fatty acids* than are present in adult fat. The two polyunsaturated fatty acids mentioned so far are linoleic acid and alpha-linolenic acid. These two fatty acids are vitamin-like because they have irreplaceable functions as building blocks in the body and because mammals cannot make them in their bodies. Hence, they have to be supplied in the diet. In fact, in the

1930-40s, linoleic acid and alpha-linolenic acid were collectively known as 'vitamin F', a term that went out of use in the 1950s.

As will be discussed in greater detail in Chapter 8, linoleic acid and alpha-linolenic acid are slowly converted to the long chain polyunsaturated fatty acids – *arachidonic acid* and *docosahexaenoic acid*, respectively, which are particularly important in phospholipids of the brain. One would not expect to find high amounts in adipose tissue of two fatty acids needed in the brain and, indeed, that is the case. However, three to four times more arachidonic acid and docosahexaenoic acid are found in body fat at birth than in adults, apparently as a reserve for later needs, especially in the brain.

Polyunsaturates represent less than 1% of the fatty acids in body fat at birth, of which docosahexaenoic acid is about 0.4%. When that very small percentage is multiplied by the normal 500 grams of fat at birth, it amounts to one gram reserve of docosahexaenoic acid. Our studies have shown that the human brain normally accumulates about 10 milligrams of docosahexaenoic acid per day during the first six months of postnatal life. Thus, if it were supplying only the brain, this one gram docosahexaenoic acid reserve would last 100 days or about three months.

We calculated that all the rest of the body needs about as much docosahexaenoic acid as the brain does so the total docosahexaenoic acid needed by the developing infant during the first six months of life is about 20 milligrams per day. If no docosahexaenoic acid was available from mother's milk or from a milk formula, and if all the docosahexaenoic acid in body fat at birth was available for growth and development, this reserve would still last an impressive 50 days.

Next to nothing is known about what triggers the deposition of fat on the developing human fetus. However, this process is known to be limited to the third trimester (Figure 4.2). Almost no fat is present on the human fetus before it is 26 weeks old, i.e. before the beginning of the third trimester. Hence the fat a baby is normally born with almost all accumulates in the final 13-14 weeks of pregnancy. Insulin is present in fetal blood before the third trimester but it does not affect glucose or fatty acid metabolism until the fetus is about 28 weeks old, the same time that fat starts to accumulate on the fetus.

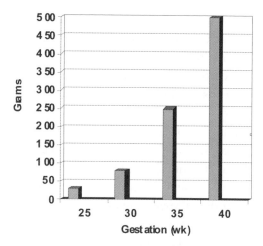

Figure 4.2. Accumulation of body fat on the human fetus during the third trimester of gestation (wk = weeks). Note that the total increase is about ten fold what is present at week 25-26. Amongst all large mammals, especially primates, this amount and rate of body fat accumulation is unprecedented.

Because of the very rapid rate of fat accumulation on the fetus during the third trimester, babies born before term (prematurely) have less fat that do those born at term. Babies born five weeks early have only about half the fat they should; those born ten weeks early have only about 15% of the fat they should. Even if pre-term babies had the same percentage of docosahexaenoic acid or arachidonic acid in their fat as term babies (a point that is unknown at present), the low body fat stores of these two long chain polyunsaturated fatty acids means that pre-term babies have lower absolute amounts of these two important fatty acids needed to help build brain membranes.

Body and Dietary Fat

Body fat in adults comes from fat in the diet and from fat the body can make. Because the body can synthesize fat, people consuming very low fat diets can still become obese. In fact, believing that low fat diets will prevent obesity is one of today's most widely held misconceptions about diet and health in affluent societies today. Consistent overeating in comparison to energy needs, no matter what the composition of the diet, leads to increasing fat accumulation. It doesn't matter whether it is bread, butter, meat, pasta, ice cream or whatever. Conversely, people around the world consume a wide range of diets but no one diet, race or culture is particularly predisposed to obesity. Obesity occurs when more dietary energy is routinely consumed than is spent for daily living, regardless of the diet.

Excess circulating glucose coming from a high intake of carbohydrates (starches, bread, pasta, rice, cereals, or sugars) can be converted to body fat at a rate *exceeding* the normal intake of dietary fat. On average, people in affluent societies eat about a ¼ of a pound or 100-120 grams of fat per day. Experiments with very low fat diets show that humans can make as much as 150 grams of fat per day. Thus, on a low fat diet the body can make more fat than one normally consumes. Therefore, the central issue in obesity is not dietary fat intake alone but how much food one eats compared to one's exercise level. Adding to the misconception that low fat diets will reduce body fat is the sad fact that intermittent or *yo-yo* dieting actually adds to the body's ability to make fat when one isn't actually dieting. This occurs because the combination of sporadic low fat diets and dieting actually stimulates the enzymes of fat production and prevents weight loss rather than the opposite and hoped for effect.

Body fat accumulates in specialized fat cells called *adipocytes*. All cells have a thin elastic membrane around them that keeps their contents inside but also lets the cell exchange molecules with the surrounding fluid. A membrane also surrounds the various working components (*organelles*) inside the cell like the nucleus, and the energy burning factories called *mitochondria*. Like other cells, adipocytes have this cellular machinery because they also have to be able to respond to messages to either store or release fatty acids.

Adipocytes are different from other cells in having their inner 'working' part pushed very much to the side to make room for the fat droplet they are storing. They end up looking rather like balloons, with the fat droplet being the air in the balloon and the nucleus and the rest of the working part of the adipocyte being the knot on the surface.

The fat droplet in a fat cell is a mixture of molecules called *triglycerides*, in which the *'tri'* refers to the three fatty acids that make up the bulk of the triglyceride, and the *glyceride* refers to the anchor or backbone to which the three fatty acids are attached (Figure 4.3). Fatty acids are molecules like metal chains but usually only with an even number of links. Most fatty acids in the diet or in the body are chains of 16 or 18 links. Fatty acids usually have a fairly rigid shape; there is some movement but the links do not move randomly in relation to each other like they can in a normal metal chain.

Some fatty acids are straight and are known as *saturated fatty acids*. Others called *monounsaturated fatty acids* have a kink near the middle that gives them a significant bend (Figure 4.3). Our bodies can make the saturated and monounsaturated fatty acids but we also get them in the form of dietary fats. The third family of fatty acids, the *polyunsaturated fatty acids*, has from two to six kinks making these fatty acids much more curved than monounsaturated fatty acids. Two of the polyunsaturated fatty acids are vitamin-like because animals including humans cannot make them like other fatty acids can be made (see Chapter 8). This means these two polyunsaturates have to be included in our diet if humans are to grow, develop and reproduce normally.

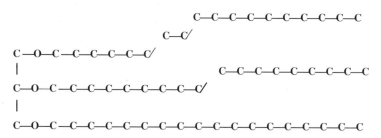

Figure 4.3. Stick model of a triglyceride with a polyunsaturated (top), monounsaturated (middle) and saturated (bottom) fatty acid. The vertical bonds connecting the three carbons on the left of the model represent glycerol. In nature, triglycerides exist with a very large number of combinations of saturated, monounsaturated or polyunsaturated fatty acids. The angled bonds represent *double bonds*, which change the shape of the fatty acid.

Dietary Fat

Whether it is from bacon, butter, salad oil, mayonnaise, fish, walnuts, or whatever, most daily intake of fat is in the form of triglycerides. During digestion, one, two, or all three of the fatty acids are broken off the triglycerides and are absorbed across the wall of the small intestine. They are then repackaged into new triglycerides in the intestinal wall and sent via the blood to the liver for further processing.

During processing of fatty acids by the liver, *lipoproteins* rich in triglyceride are formed and secreted into the blood. It is as lipoproteins that most lipids circulate in the blood. It is possible, but very unlikely, that after digestion the same three fatty acids could reassemble back on the same glyceride anchor thus making up a triglyceride molecule exactly the same as it was before it was digested. Hence, the intestines themselves and the liver do a large amount of further recombining of fatty acids to make new triglycerides and phospholipids, which circulate in the blood.

The fatty acids that make up the triglycerides of body fat come from a combination of the fatty acids that were eaten and those made in the body. Analysis of a tiny sample of body fat therefore gives a good indication of a person's choice of dietary fat over the previous months. If one never previously ate fish, this would be obvious in body fat from the lack of *long chain omega-3 polyunsaturated fatty acids* that are mostly peculiar to fish and seafood. If one then started eating fish, it would become evident within three to four days from the changed fatty acid pattern in fat. If one had always eaten some fish and then completely stopped, this would eventually become measurable but it would take months before this was clearly obvious because fatty acids in body fat are used up rather slowly.

Experiments in which changes in body fat composition are measured after changing dietary fat intake demonstrate that the fatty acids in adipocytes do not stay there forever; the fatty acids in body fat are slowly and continuously changing regardless of whether one changes what one eats. This release and replacement or *turnover* of fatty acids is regulated by hormones and by stimulation from the nerves connected to fat cells.

All adipocytes have a blood supply providing them with oxygen, nutrients, and the fatty acids available for accumulation. The blood also takes away the carbon dioxide, wastes, and the fatty acids released by adipocytes. Hormones influence the speed of these processes and arrive at the adipocyte via the blood.

Insulin is one of the most important hormones affecting release or uptake of fatty acids by adipocytes. Higher insulin levels (such as occur after a meal) increase fatty acid uptake into adipocytes and inhibit their release. When insulin decreases two to three hours after the last meal or on waking, this removes the brake on the outflow of fatty acids from adipocytes and they are again released into the blood. These released fatty acids are fuels when glucose is low. Other hormones, particularly those involved in the stress response like adrenaline, can stimulate adipocytes to release more fatty acids because of an impending need for additional fuel to aid in the reaction to a stressful or exciting situation.

Phospholipids

Humans or animals with no visible body fat still have chemically measurable fat in them because the membranes surrounding and within each cell are made not only of proteins but also of two fatty substances called *phospholipids* and *cholesterol*. In one way, phospholipids are chemically similar to triglycerides because they are both made principally of fatty acids. However, there are five important ways that phospholipids are very different from triglycerides (Table 4.2):

Table 4.2. Significant chemical differences between phospholipids and triglycerides that markedly differentiate both their chemical nature and their biological roles.

1. Phospholipids contain phosphorus but triglycerides do not.
2. Phospholipids are components of membranes and, unlike triglycerides, are not stored as droplets in the cell.
3. Phospholipids contain two fatty acids instead of the three in triglycerides.
4. Phospholipids are richer in polyunsaturated fatty acids, especially the long chain polyunsaturates, arachidonic acid and docosahexaenoic acid. Triglycerides are richer in saturates and monounsaturates, and usually have very low amounts of long chain polyunsaturates.
5. Unlike triglycerides, which are *storage fats* used for energy, phospholipids are *structural fats* used to maintain cell form and function.

Two key features making phospholipids very important components of cell membranes are the several different possible head groups and the high proportion of polyunsaturated fatty acids in phospholipids. Phospholipids are especially important in organs such as the heart, brain and eye, each of which depends on relatively high electrical activity.

Just exactly why phospholipids are so important in cell membranes is under intense investigation. In part, these molecules help docking points *(receptors)* on the surface of the cell recognize and transport substances like glucose that are needed inside the cell. Other receptors anchor hormones like insulin that influence how the cell works. Phospholipids help orient a myriad of proteins in appropriate places and configurations in the cell's membranes. They are also part of a cascade of *signalling molecules* within the cell that direct the cell's response to virtually all stimuli received and reacted to by the cell.

The chemistry and molecular biology of cell-cell interactions is immensely complicated and growing the more so as new technologies expand the tools for their investigation. Suffice it to say at the moment that phospholipids and their particular fatty acid profiles are an integral part of these cellular mechanisms.

Structural Lipids and Storage Fats

Distinguishing between triglycerides and phospholipids is important in understanding why lipids are biologically important, especially for human brain evolution. The overall term, *lipid* defines these 'fatty molecules' on the basis of their common chemistry, which is that they are extractable from tissues or foods by organic solvents like chloroform. Hence, lipids include all phospholipids, cholesterol, triglycerides, as well as other related substances.

The common term – *fats* – refers to fatty acid-rich lipids that are solid (fats) or liquid (oils) at room temperature. Thus all fats or oils are triglycerides and are also lipids. However, because substances like cholesterol are extractable with organic solvents but do not contain fatty acids, not all lipids are fats. These definitions overlap and are admittedly confusing. Regardless of the ambiguity between chemical and biological terminology, the important thing is that the storage fats and structural lipids are functionally quite distinct.

The confusion is more than with overlapping definitions because when all the fats (lipids) are extracted from tissues by organic solvents, the phospholipids, cholesterol and triglycerides all come out together. It is rather like with apples and oranges; they are different to feel, look at or to taste, yet they are both sweet, have a peel and have seeds inside. Once squeezed and the juices combined, it is no longer possible to tell how much of each contributed to the total.

So it is with phospholipids and triglycerides; chemists lump them as fats and when they are extracted from a food sample or tissue, they can no longer be told apart unless further sophisticated analysis is done. On the other hand, biologists recognize that phospholipids are structural lipids occurring only in cells membranes while triglycerides are storage fats almost always found in droplets and from which cells can derive energy. Like phospholipids, cholesterol is a key structural lipid and cannot be used as a fuel.

After extraction with an organic solvent, brain tissue is chemically about half fat, but this fat is not 'fat' in the biological sense because almost none of it is in the form of fat droplets, i.e. it is not triglyceride. Rather, the brain's fat content is almost exclusively in phospholipid and cholesterol and is all in the brain's cell membranes. Phospholipids and cholesterol help insulate the main wiring of nerve cells so that messages go where they should. These two lipids also help constitute the billions of finely organized branches and synapses between nerve cells. Thus, essentially all the fat in the brain is *structural lipid*.

Other organs like muscle or liver also have structural lipids in cell membranes but, unlike the brain, they also store some triglyceride in fat droplets. Hence, part of their fat is structural lipid in membranes and part is storage fat. Adipocytes also have phospholipids in their membranes but they are unique in having a very large storage droplet of triglyceride as well.

When measured chemically, even the leanest animals always contain 4-5 % fat. In such lean animals, this extractable fat is comprised almost entirely of phospholipids and cholesterol in the cell membranes of every organ and tissue in the body. When opened up for dissection, these lean animals have few visible fat deposits as such. Thus, the leanest animals physically have very little body fat. In practice, most animals usually have low to moderate amounts of true storage fat in subcutaneous or visceral depots. The amount of storage fat varies with several factors, most notably the season and the food supply. Therefore, adding the structural lipid and storage fat together gives a value of chemically extractable total fat of something usually between 5-10% of body weight.

Some animals, especially marine mammals like seals living in cold water, have large amounts of true storage *subcutaneous* fat. Hence, they can be as fat as obese humans. In that sense, human body fat is not unique in the animal kingdom. However, besides humans, no animals living on land truly have fat *babies*, e.g. with noticeable fat deposits under their skin. Some marine animals have fat babies but humans seem to be unique amongst land-based animals in this respect.

Fat Metabolism, Ketogenesis and Brain Fuel

A fatty acid that has been burned as a fuel is said to have been *oxidized.* *Beta-oxidation* is the more correct name because there are various ways oxidation of fatty acids takes place but only beta-oxidation leads to fuel production. Rancidity of fats occurs because of *peroxidation* of fatty acids but this is a very different process than the oxidation of a fatty acid for fuel. Fatty acid oxidation takes place in mitochondria.

The main signal telling mitochondria to oxidize fatty acids is low insulin in the blood, such as would occur after overnight fasting. Low blood insulin tells adipocytes to break individual fatty acids off the triglycerides in their fat droplet and to release the resulting *free fatty acids* into the blood. Cells that can oxidize fatty acids then take them up from the blood and transfer them to their mitochondria where the fatty acid will be broken down two carbons at a time. These two carbon units are used to make *acetyl CoA,* which enters the final process of generating ATP. Carbon dioxide and water are the final waste products of this process.

Not all cells can burn fatty acids as fuels. Hence, when blood glucose is low and muscle or heart cells are happily burning fatty acids as a substitute, the brain still burns some glucose but also counts more and more on a different alternative fuel – *ketone bodies* (ketones for short). Ketones generally have a bad reputation because they are associated with uncontrolled insulin-dependent diabetes. Hence, high blood ketone levels (above 20 millimolar) are a sign of serious metabolic disturbance. However, the mild to moderate blood ketone levels produced during normal (or even prolonged fasting), or which are normally present during infancy, rarely exceed 2-3 millimolar and are entirely benign. In fact, to the contrary, mildly elevated blood ketones are *essential* for normal brain development, at least in humans.

In the process of *ketogenesis*, two of the accumulating acetyl CoAs condense together and form the first ketone – *acetoacetate* (Figure 3.2). Acetoacetate has four carbons and can be converted to a more stable ketone, *beta-hydroxybuyrate*, also with four carbons. Acetoacetate can also lose one of its carbons and become the third ketone, *acetone*, which is the same substance that is a household solvent commonly used for removing nail polish. Acetoacetate, beta-hydroxybutyrate and acetone are all released into the blood from the liver where most of the body's supply of ketones is produced. Organs like the brain readily use ketones when blood glucose starts to drop.

Humans have a better than average ability to raise ketone levels in the blood. My hypothesis is that evolution of a greater capacity to make ketones as well as to build up fatty acid reserves in fat stores in order to make the ketones arose during or before human brain expansion. In all species, the brain is a sophisticated organ and is very vulnerable to its energy supply. To both expand brain size and increase its sophistication so remarkably as occurred during human evolution would in all likelihood have required a reliable and copious reserve of dietary energy for at least a million years. Glycogen stores cannot serve this fuel function for long because they are two small relative to the energy needs of large-brained species.

Hence, some hominids began to evolve larger brains as a consequence of acquiring a more secure and abundant food supply that helped build up more long-lasting body fat stores. Evolving a larger brain depended on simultaneously or previously evolving significant fat stores and having reliable and rapid access to the fuel in those stores. Although fat stores were necessary, they were still not sufficient without an improved capacity for ketogenesis. This combination of outstanding, long term neonatal fuel store as well as rapid access to the fuel was the key insurance for expanding the hominid brain, insurance that was not available to other land-based neonates, especially other primates.

Fatty Acid Oxidation and Synthesis in Infants

Acetyl CoA is an important molecule derived mainly from glucose, fatty acids or ketone body oxidation. When mitochondria are replenishing the cellular fuel, ATP, normally the cell uses the available acetyl CoA only to make ATP. However, once ATP is topped up, acetyl CoA can be used to build many different molecules, like new fatty acids or cholesterol that are then incorporated into cell membrane lipids. If the body's main need at the moment is to make more fuel because blood glucose is low, then the available acetyl CoA will be used as a fuel and there is little chance that it will be used to make a new fatty acid instead.

Thus, the normal situation in fasted (energy-depleted) *adults* of all animal species is that acetyl CoA is primarily used as a fuel. Acetyl CoA has a myriad of potential uses so it makes sense to prevent it from being used to build new molecules when what the body really needs at that moment is more fuel for energy-starved cells. In contrast to the fasting situation, when there is sufficient glucose around, then excess acetyl CoA is channeled towards building tissue protein or fats.

That is the normal situation in adults but infants use acetyl CoA differently than do adults; infants can synthesize new fatty acids while *simultaneously* burning other fatty acids as fuels. Hence the acetyl CoA that adults would only channel into ATP synthesis can be diverted by infants towards synthesis of new fatty acids. Hence, unlike in adults, fatty acid oxidation in human infants isn't strictly limited to the goal of making more fuel.

Babies are usually fed sufficiently frequently that ketone production would never be stimulated if it depended only on being hungry, yet babies constantly have higher blood ketones than adults. This is the normal situation and is true regardless of whether blood ketones are measured before or after feeding. Furthermore, fat stores normally expand during infancy so, despite constantly producing ketones, some fatty acid synthesis and storage is also going on simultaneously.

The fact that babies tend to have higher blood ketones than adults is all the more interesting considering that their brains are proportionally much bigger than in adults and that their brains are able to extract ketones from the blood three to four times more effectively than in adults. In other words, all things being equal, it would take three to four times more ketones to be pumped out by the liver just to keep the same blood ketone level in infants as in adults because the infant brain can remove the ketones that much more effectively. That means the infant must be producing more ketones or using them more slowly outside the brain than occurs in adults. Since blood levels of ketones are actually higher in infants despite their brain's greater ability to take them up from the blood, their net ketone production compared to utilization is considerably greater than in adults.

Ketones and Lipid Synthesis

Ketones are important alternate fuels to glucose for the brain but that is not all they do. Ketones are versatile molecules and are used as building blocks, for instance, to make cholesterol and saturated fatty acids. Thirty years of studies done in rodents, non-human primates and humans were the foundation for this fundamentally important observation.

On the surface, this situation seems counterproductive. Why recycle an oxidized fatty acid through ketone bodies into a new fatty acid or into cholesterol for that matter? Where does this happen and why? The brain definitely can use ketone bodies as building blocks for structural lipids for myelin and nerve endings. However, other organs like the intestines, lung and liver also accumulate lipids made by reutilizing oxidized fatty acids to make new fatty acids or cholesterol.

Compared to glucose, ketones are actually preferred sources of carbon to make fatty acids or cholesterol. The infant brain has a very high demand for cholesterol, but it can't import sufficient cholesterol from the rest of the body. Hence, the ability of infants to turn oxidized fatty acids into brain cholesterol via ketones starts to make sense; this process is intimately linked to supporting brain development.

The issue during human brain evolution was basically one of meeting the requirements of both brain development and high brain energy demands while simultaneously expanding the brain. This had to be done while using small molecules like ketones that have mixed uses, i.e. that are good fuels but, simultaneously, also are good molecules to build membrane lipids.

One potentially relevant point is that the brain is not very good at burning fatty acids as fuels. This is good news because the brain has no triglyceride stores, so if it were able to oxidize its own fatty acids and use them as fuels, it would have to use up its structural fatty acids in membrane phospholipids. Every time this happened, a new period of membrane reconstruction would be needed to restore membrane structure and brain function.

An analogy is the inability of the liver to make ketones while simultaneously using them for energy; if the liver could oxidize ketones, its high metabolic rate would consume all the ketones it produces and there wouldn't be enough left over to export to other organs, especially the brain. Seen in this light it makes sense that the brain doesn't burn its own fatty acids to any significant extent. Metabolically expensive organs have to clearly separate molecules they use as fuels from molecules they need in membrane structure.

Not all fatty acids are equally easily oxidized and converted to ketones. Interestingly, the two vitamin-like polyunsaturated fatty acids (linoleic acid and alpha-linolenic acid) appear to be the preferred fatty acids used to make ketones. Linoleic and alpha-linolenic acids are thought to be important as direct precursors to other longer chain polyunsaturates needed in membranes so it seems paradoxical that they would also be the most easily oxidized of the common dietary fatty acids.

Actually, this is no more paradoxical than using ketones both as fuels and as building blocks for brain lipids; it just means that these two fatty acids serve different and sometimes competing functions. Dietary supplies of linoleic acid and alpha-linolenic acid are normally fairly generous and humans rarely undergo extended periods of voluntary fasting with prolonged ketone production, so the use of these two fatty acids for ketone production rarely risks inadequate amounts remaining for synthesis of long chain polyunsaturates.

Fat Babies and Child Birth

Despite the merits of body fat stores for brain development in human infants, as Elaine Morgan pointed out, the larger brain and extra fat have an important disadvantage - they make babies bigger and therefore their birth more difficult. The birth canal passes through an oval opening in the base of the pelvis. The pelvis is like a basket with a large hole in the bottom. Inside the basket are the lower abdominal organs including the large intestine and, in the female, the reproductive organs. During pregnancy, the enlarging uterus with its fetus and placenta become occupants of this space as well. These organs (and the developing fetus) are kept from falling through the hole in the base of the pelvis by the support of strong muscles surrounding the anus.

A double challenge for the pelvic area affected childbirth in pre-human hominids and humans. First, it had to provide a sufficiently large opening for the much larger human baby during birth. Second, the pelvic opening had to be kept small enough to support walking and running upright, while also preventing prolapse of the lower abdominal organs out through this opening. Hominids didn't just evolve large-brained babies with lots of body fat; they also evolved bipedalism.

The main part of the back and lower human pelvis supports the vertical spine and anchors the legs. The lower front part of the pelvis completes the oval-shaped basket and gives strength to the hip joints. It also creates the opening of the birth canal. The front bony part of the pelvis, or pubis, has an expandable joint in the middle. This joint is hard and essentially immovable most of the time but softens and becomes stretchable just in time for childbirth. Childbirth in humans can be a difficult process for both the mother and the baby but this appears not to be so in other primates.

The main reason that birth seems to be less difficult in other primates is that, compared to humans, the pelvic opening is considerably larger relative to the infant's head size and shoulder width. Humans seemed to have pushed fetal brain and body size to the maximum possible limit allowed by the pelvic opening. In so doing, childbirth is commonly painful if not outright difficult for most women. In some women, childbirth is clearly traumatic and risks harming the baby.

The whole organism has to benefit from (or at least cope effectively with) the morphological changes being introduced. The advantage of a hominid woman's new ability to walk on two legs had to accommodate not only the increased difficulty with supporting her abdominal organs that were now vertically stacked, but also had to accommodate the birth of larger-headed, fatter infants. The bigger brain and the added body fat constitute, at least in modern times, a risk to safe, trouble free childbirth. These risks were evidently not sufficient to have prevented evolution of the big-brained fat human baby.

Nutritional status of the mother is one of the important factors affecting the difficulty of childbirth. A woman who is inadequately nourished during her own childhood and adolescence will usually be smaller than average and will have a smaller than average pelvis. She may get pregnant before she has finished growing and so will have to share the nutrients she consumes with her developing baby. She may have a normal or a small baby depending on the type and timing of the nutritional compromise she still faces. Thus, being small may not necessarily make childbirth more difficult if the baby is also proportionately small. However, if the mother is small but the baby is closer to normal size, childbirth is bound to be more difficult. Unless maternal malnutrition is prolonged and severe, the baby is more likely to be closer to normal size because its nutrient supply is favoured during pregnancy.

The point is that maternal nutrition influences all the parameters relevant to child birth – the size of the mother's pelvic opening, the amount of fat on her baby, and her baby's body and brain size. Evolution of human fetal fat occurred in a genetic and nutritional environment that also favoured brain expansion but which, at the same time, had to ensure an adequate opening in the maternal pelvis.

Prematurity, Low Birth Weight and Fat Stores

During the first two thirds of gestation, the human fetus is lean like the fetuses of other terrestrial species, i.e. it has essentially no body fat deposits. When the human fetus is 25 to 26 weeks old, some cells under the skin start to differentiate from connective tissue and become fat cells or *adipocytes*. The differentiation of adipocytes is the start of a process during which body fat content will more than *double* during the next 6 to 7 weeks and will more than *triple* during the next and final 6 to 7 weeks of pregnancy.

As a result, about 500 grams (about 1¼ pounds) of fat will normally be deposited on the body of the fetus by the time it is born. This fat forms a more or less uniform layer under the skin of the trunk, face and limbs. Its distribution is unlike that of fat in the obese adult or adolescent, in particular because almost none of the fat of the newborn surrounds the organs inside the abdominal cavity where, in adults, it is associated with a higher risk of diabetes.

Adipocyte number increases two to three fold and their fat content increases up to fourteen fold during the first two to three post-natal years. As a result, during the first year of life, total body fat content increases about four fold and parallels body weight gain. Adipocytes have an enormous capacity to expand in size so the total increase in fat during the first few postnatal years can be accommodated by adipocytes present at birth rather than requiring differentiation of new adipocytes. However, during childhood and adolescence, total body fat content increases more than the adipocytes formed during infancy can accommodate, so further adipocyte differentiation usually occurs during this later period.

Nutritionists and pediatricians have always been concerned about protein intake and lean tissue growth as markers of health in the fetus or infant. However, fat deposition is also a key aspect of healthy human fetal development in the third trimester. In fact, during the final weeks of the third trimester, Fred Battaglia (University of Colorado) showed many years ago that the near-term fetus spends considerably more energy on fat deposition than it does on protein synthesis and lean tissue growth. Fat deposition is by far the major component of the increase in body weight during the third trimester. This fat accumulation is desirable, indeed *necessary,* for healthy post-natal development in humans.

Low body fat is the most important contributor to low birth weight in humans. Fetal body fat is deposited primarily during the third trimester so premature birth is associated with a disproportionately low amount of body fat. Premature infants risk compromised organ function because of underdeveloped digestive and immune systems and inadequate fat reserves. This is a problem similar to that faced in malnutrition when energy reserves are low to nonexistent and poor sanitation often results in gastrointestinal infection and poor nutrient absorption. In effect, even the healthiest premature infant is at higher risk of malnutrition than is a healthy term infant. Modern hospital care decreases this risk, which would otherwise more often be fatal.

Premature infants have several problems linked to energy metabolism. First, they have low fat stores because they are born before fat stores have been adequately deposited. Second, because they have less body fat, premature infants have proportionally larger brain size even than healthy term infants. Third, small bodies have proportionally larger surface area, which increases heat loss and hence energy demand. Effectively, the premature infant faces the challenge of simultaneously fuelling the high cost of staying warm, developing the brain and other tissues, but without the fuel reserves of the term infant.

This problem is more acute the more premature the infant. Birth weights under one thousand grams (birth at 22-25 weeks, compared to the normal 39-40 weeks gestation) are not uncommon today but this is an extremely unhealthy situation. In most cases, pediatric intensive care is needed not only to try to maintain normal growth but to reduce unnecessary fuel demands so as to provide a better chance of normal brain development.

In the face of limited glucose and poor alternative energy sources, prematurity, low birth weight or malnutrition after birth each stunt brain development and growth. This is because the body may actually be obliged to choose between meeting the high energy costs of maintaining body temperature versus development of organs, especially the brain. Normal physical growth and psychomotor development are the best evidence that the brain's essential fuel needs are being met and that nutrients are left over for tissue expansion and growth.

Fat stores are more critical to energy needs during early development in species with larger brains. This is the central feature of the uniqueness of human brain evolution. Body fat is the best reservoir of an alternate energy source to glucose to fuel the expensive brain between meals. Even expansion of body fat stores is energy consuming. It is soon after birth that the development of body fat stores is most vulnerable to inadequate nutrition, which is precisely when humans have the most need of body fat.

Low fat reserves in premature infants creates competition to oxidize for energy the very long chain fatty acids that should be destined for synaptic membranes in the developing brain, and represents a huge challenge for normal development. Low fat reserves represent a similar limitation to brain development in other species, including non-human primates, and prevent them from evolving human-type brain size and function. In effect, all other species are born with fuel reserves equivalent at best to premature human infants. Some are born helpless like humans; others are born 'fully functional' and can identify their mother and feed immediately. Regardless of their maturity, they are born without clear fat reserves underlying the skin. Those born precocial have brains that are better developed, which allows them to function at least physically.

More importantly, though, most other species are not expanding their brain function significantly after birth; the energy investment in brain function is smaller because the brain is proportionally smaller than in humans and because much of the brain's maturation was completed before birth. Hence, it would be a waste to develop postnatal fat stores because the brain isn't big enough and because further functional capacity isn't needed. The piglet is a good example of such a precocial, highly functional species at birth, but it cannot develop sufficient ketosis to save its own life if starved for more than a day or two. In the context of survival, it has a high functioning brain at birth but that brain function will not expand much postnatally because there are no fuel reserves to support it.

***

Healthy newborns do not lose their body fat as they grow; in fact, they gain more during the first few years of life. In effect, after birth they continue to build fat stores to accumulate fuel insurance for their brains. One might well ask how this reserve could be important if it expands rather than being used up during early development. Indeed, why do fat reserves develop so rapidly during the third trimester when the fetus has no real use of its brain, at least for survival?

Body fat expands during late fetal development essentially to provide the best possible *insurance* that postnatal development will proceed uneventfully despite inevitable nutritional 'bumps in the road'; nobody guarantees your food supply. One is led to conclude that for so much metabolic and genetic effort to be invested in the brain during fetal development while the organism is extremely vulnerable and has no real use of the rapidly developing but still very immature cognitive capability, this investment must happen under conditions in which the brain is highly protected. In other words, the brain's vulnerability is well masked.

Chapter 5

Nutrition:
The Key to Normal Brain Development

A changing environment creates *selection pressures* that are an important stimulus for evolution. Most major environmental change such as volcanoes, glaciation, or rifting in the earth's crust involves climate change affecting temperature and the availability of food or fresh water. Abundance of food or water could rise or fall with the climate change and also create selection pressures. How the pressure works depends on the species involved, i.e. it could favour survival of some while straining survival of others. To survive, either a sufficient number of individuals within a species is capable of adapting to the environmental change or must be capable of moving away from it, probably to face new challenges.

Those individuals that survive and reproduce do so because they are already sufficiently adapted, i.e. different, to be able to tolerate the change. That adaptation, whether it is beak shape, rudimentary standing ability or whatever, then becomes the focus for the distinctiveness from the parent population that is the seed of the new species. The concept of selection pressure can have a negative connotation in the sense that the environmental change is seen as exposing survival limitations; only a few individuals are sufficiently different within the species to both be members of that species but also to survive and reproduce in the changing new environment.

In my view, changing environments not only force evolution, they also *permit* it. A changing environment may unlock a certain genetic potential for morphological or physiological change that previously remained silent. Some individuals within a species and certainly some species as a whole may be better prepared for and better able to take advantage of a particular climatic or environmental change than others. Genetic or physiological preparedness means that a change in the environment is permissive; the changing environment doesn't necessarily force change, it allows it.

My hypothesis is that key nutritional changes in a new environment permitted human brain evolution because the brains of some hominid species were genetically prepared for such change. There was no selection pressure. Nothing forced the brain to become larger. Nothing forced babies to develop body fat. Hominid survival was not at risk; several survived for million year periods without substantial change in brain size.

In fact, inadequate human diets today are evidence that by exposing the brain's developmental vulnerability, evolution in hominids actually increased the risk of not surviving. As we saw in Chapter 3 and will discuss again in the next several chapters, cognitive advancement can still today be victimized by a developmental problem in the brain. If vulnerability to nutritionally imposed developmental delay remains a paramount feature of the human brain today, the risk of that developmental delay must have been essentially non-existent for it to have been retained as such a dominant feature during human brain evolution.

The increasing vulnerability inherent in the process of enlarging the hominid brain demonstrates that brain enlargement and increasing cognitive potential were neither evolutionary objectives in themselves nor were they solutions to any problem of survival. The fact that increased developmental vulnerability of the brain was tolerable, i.e. did not substantially impair reproduction, survival or, indeed, evolution of the Australopithecines into *Homo*, means that the organism as a whole was not put at substantially greater risk by increased vulnerability of the enlarging brain. That, in turn, must mean that food and habitat of the species became increasingly secure as brain expansion occurred.

***

Primates, especially chimpanzees, are intelligent compared to other mammals. They occupy stable nutritional niches and have established social order and communities. In many ways, they are genetically nearly identical, but behaviourally simplified, versions of humans. However, one of their important physical differences with humans is in jaw and dental structure. Flatter molars with larger surface area occur in species consuming predominantly plant-based diets that require a lot of chewing to degrade the fibrous support structures of the shoots, leaves or roots prior to digestion. The grinding of food also requires a larger, heavier jaw and bigger jaw muscles.

If diet can so markedly affect jaw and tooth shape, it can affect other physiological and biochemical attributes as well. Differences in diet between humans and chimpanzees represent important differences in nutrient intake. Given the susceptibility of the human and non-human primate brain to function suboptimally when fuel (energy) and certain nutrients are consumed in lower than necessary amounts, it is important to explore the relation between nutrition, brain development and brain evolution. Adverse nutritional, disease, economic, and psychosocial circumstances expose developmental vulnerabilities preventing at least one quarter of the world's population from attaining the high level of brain function of which humans are theoretically capable.

Maternal Nutrition, Prematurity and Early Development

Normal growth of the mammalian fetus depends on the mother having adequate nutrients and energy during pregnancy. *Lower birth weight* of the infant is a common consequence of intrauterine growth retardation and maternal undernutrition during pregnancy. Maternal nutrition sufficiently bad to cause low birth weight or prematurity (less than 37 weeks gestation) is frequently associated with stunted intellectual capacity at school and lifelong mental disability of the offspring. The lower the birth weight, the higher the risk of handicap, including mental retardation, visual impairment with retinopathy and blindness, hearing impairment, cerebral palsy, spasticity and autism. The prevalence of these developmental handicaps in infants of normal birth weight is under 1%. However, they occur in 20% of babies weighing under 1500 grams at birth, i.e. the risk is 20 fold higher than at normal birth weight. Nowadays, medical advancements ensure that very low birth weight babies survive longer after discharge from hospital, but the occurrence of handicap in these children is not changing. Thus, their perinatal mortality is decreasing but their long-term risk of incapacity is not.

Michael Crawford of London Metropolitan University in the UK has extensively researched the nutritional basis for poor neurological development in low birth weight and premature infants. His work points to a lack of several vitamins and minerals in the maternal diet as being associated with low birth weight. He has noted the difficulty that premature and low birth weight infants have in meeting their requirement for the key long chain polyunsaturated fatty acids such as docosahexaenoic acid during early neurological and visual development.

Crawford was also the first to recognize the parallel requirement for long chain polyunsaturates in both vascular and neurological development. The importance of an intact vascular system to normal neurological development is apparent from the high incidence of intracranial hemorrhage in infants with impaired neurodevelopment, and from the high incidence of placental infarction causing poor fetal development. Part of the challenge facing brain development is adequacy of the blood supply to the brain; when the capillary network in the brain itself or in the placenta is also underdeveloped and fragile, the brain's development is at much higher risk.

Clear evidence for long-term neurodevelopmental problems arising in very low birth weight babies comes from Maureen Hack's work in Cleveland first published about a decade ago. She studied the development of very low birth weight infants that, by definition, weigh less than 1500 grams at birth. Her group focused on those low birth weight infants that had growth problems around the time of birth.

The average birth weight of the children she studied was actually about 1,200 grams and they were born 10-11 weeks premature. Her babies faced an incredible challenge, in that just to grow to the equivalent weight of normal term infants at birth (3,600 grams), they had to triple in size but without the benefit of being in the uterus for the last quarter of pregnancy. Aside from prematurity, they had no confounding diseases; thus, they were as healthy as very premature infants can be.

Clinical and socioeconomic circumstances were similar in the infants with small head size compared to children with normal head size.

Twelve per cent of the children in Hack's very low birth weight group had abnormally small head size at birth. Twice as many had small head size 10 weeks later when they reached normal term, and 13 % still had small head size at eight months old. Hence, despite the best available medical care, once born, it is very difficult for the premature infant to attain normal growth and development.

Hack's group found that, independent of all other factors, normal head size at eight months old in these infants was the best predictor of neurological and behavioural development at three years old. Notwithstanding some confounding effects, head circumference during the first year of life is a good measure of quantitative features of normal neurological development, i.e. the size, volume, cell density, and DNA and lipid content of the brain.

Using the same cohort of children but studying them at three years old, Hack compared the performance on normal tests of neurodevelopment at school once they reached seven to eight years old. On most measures those with small head size at eight months old were much more likely to be low achievers. These tests assessed language development, hand-eye coordination, social competence, behaviour, and achievement at school.

They concluded that poor school age performance is almost guaranteed unless an infant with a small brain at birth has the chance to catch up to normal by the time it is eight months old. If catch up in growth is to be of any help, it has to occur within the first year of life. Otherwise children and adolescents that were malnourished during infancy will continue to have small head size and poor neurological and behavioural outcomes. Catch up requires improvement not only in nutrition and energy intake but also good psychosocial support.

As Michael Crawford has frequently noted, the problem of prematurity and low birth weight is no longer rare. Not only are smaller and smaller premature infants 'technically surviving' as medical technology improves but the proportion of total births represented by small babies is increasing in affluent and developing countries alike. The cause of this growing problem is multifactorial: Poor maternal socioeconomic circumstances play a key role in poor diet selection and higher risk of nutrient deficiencies in the fetus.

In fact, Crawford's work points to poor nutrition having a negative impact on pregnancy outcome before conception. Poor nutrition reduces the size of follicle development in the ovary thereby affecting the earliest stages in reproduction. This makes it difficult if not impossible to correct maternal nutritional deficits once pregnancy is established. The most viable public health option is to more vigorously pursue the *pre-conceptional* prevention of low birth weight by improving the health and nutrition of the mother before pregnancy.

Malnutrition and Brain Development

The adult brain is spared during malnutrition, e.g. its function is maintained and it loses the least weight compared to those organs occupying the majority of body weight. This *metabolic sparing* of the brain is a feature of the physiological response to undernutrition or starvation in all mammals, and was probably an early and essential feature of mammalian brain evolution. Thus, the evolution of efficient protective mechanisms to meet the brain's energy requirement leads directly to its sparing during malnutrition but masks the underlying vulnerability of brain function.

Because of their rapid growth and immaturity, children are the most adversely affected by hunger and malnutrition. Unlike in adults, the brains of young children are still very vulnerable to starvation. This is due to the proportionally higher energy requirements of the brain during childhood than at any other time later in life. Severe malnutrition rapidly dampens the exploratory behaviour of children and leads to apathy. Brain activity is metabolically expensive so less responsiveness and reduced physical activity allows the developing brain to better withstand the severe metabolic impact of malnutrition, but at the risk of long-term suboptimal performance.

The primary deficit in the brain that is caused by nutritional deprivation is smaller brain size, whether in relation to age or to body weight. The reduction in brain size is accompanied by a reduction in weight, volume, number of cells, and total amount of DNA and lipid in the brain as a whole.

Different timing of the brain growth spurt in different species must be taken into account in modeling this process because the same nutritional deprivation induced after the *brain growth spurt* has much less impact compared to during the brain growth spurt. The young of a species with a prenatal brain growth spurt, e.g. the guinea pig, are less affected by postnatal malnutrition or by excess competition for milk caused by a large litter than are the young of a species whose brain growth spurt is perinatal, e.g. the human, or whose brain growth spurt is postnatal, e.g. the rat.

In addition to impairing protein synthesis, malnutrition in human infants less than a year old also reduces the total content of cholesterol and phospholipids in the brain. Brain DNA levels (a marker of brain cell division) are relatively unaffected unless malnutrition is prolonged into the second year of life. These changes parallel those observed in animal models and suggest that the impact of severe malnutrition on the developing brain has a broadly equal and severe effect on lipids as on protein and DNA synthesis.

Malnutrition at birth reduces brain cell number and size. Malnutrition starting later in infancy may only affect brain cell size because normal brain cell number should have been achieved by then, even though these cells are impeded from achieving their full growth potential. In the latter case involving normal cell number but small size, is it possible for refeeding to restore some normality to brain development. However, if malnutrition occurs earlier, i.e. before the critical period for establishing brain cell number has passed, the deficit cannot be corrected and brain function will be permanently impaired.

Malnutrition, Brain Lipids and Myelination

Malnutrition during the suckling period impairs myelination. This impairment arises because malnutrition prevents development of a full compliment of myelin-producing glia and because they lack the necessary amount of substrate needed to make the normal amount of myelin. Impaired brain cholesterol synthesis is a key part of the problem because cholesterol is an important part of myelin. In addition to cholesterol, myelin lipids also contain large amounts of very long chain saturated and monounsaturated fatty acids that are relatively unique to the brain. As with cholesterol, these fatty acids are almost entirely made within the brain. Their synthesis is impaired during malnutrition because the carbon that would have been used to make them gets used as a fuel.

During malnutrition, the deficit in myelin lipid production arises in part because normal rates of cholesterol or long chain fatty acid synthesis within the brain cannot be achieved during malnutrition. It also occurs because very little lipid gets into the brain from outside it to make up for what is being made too slowly within the brain. Malnutrition redirects the glucose (or ketones) that would have been used for cholesterol and long chain fatty acid synthesis towards fuels to keep organs functioning, especially the energetically expensive ones like the brain, kidneys and heart. Malnutrition during the critical period of myelination leaves less substrate for protein, long chain fatty acid and cholesterol synthesis needed to make myelin.

In effect, malnutrition forces all systems of the body to compete in the face of fuel deficit and some processes, like myelin lipid synthesis, simply cannot compete and be sustained at the normal rate. Thus, less total myelin is produced and it is thinner, reducing the fidelity of message transfer from neuron to neuron. The timing and duration of malnutrition determine which areas of the brain undergo a deficit or distortion in myelination because myelination is completed at different times depending on the nerve tract and brain region involved.

Malnutrition versus Specific Nutrient Deficiencies

Malnutrition occurs when total food intake (energy, protein and required nutrients) is below the requirement level for a period of several weeks or more, resulting in weight loss as well as poor growth in children. There is an important distinction between generalized malnutrition versus a diet providing adequate energy and protein but inadequately supplying one or more specific nutrients needed in the diet: In the first case, malnutrition and starvation due to insufficient food intake and poor sanitation are widespread in the world today. By impairing nutrient absorption or increasing disease, poor hygiene exacerbates malnutrition and *vice versa*. In the second case, despite sufficient energy and protein intake, moderate to severe deficiency of a nutrient like vitamin C still causes the specific symptoms of scurvy.

Malnutrition impairs growth and learning in all animals, not just humans. Unless it is severely deficient, an iodine deficient diet does not impair growth but it does impair learning in animals and humans. Hence, iodine deficiency can be part of malnutrition but on its own, iodine deficiency has a more limited and specific effect than that of generalized malnutrition. Species possessing more advanced brains are more severely affected by iodine deficiency. Specific nutrient deficits are therefore not the same as general malnutrition and are more severe when the attribute affected is better developed or more dependent on that nutrient, e.g. brain function in iodine deficiency.

Malnutrition clearly damages brain development on several fronts. Specific nutrient deficiencies can also be very damaging even when total dietary energy and protein intake is sufficient. For example, copper deficiency causes tremor and neuromuscular instability because it induces *hypomyelination*. This occurs because of a specific impairment in the ability to convert saturated to unsaturated long chain fatty acids used in myelin lipid synthesis. A deficit in availability of polyunsaturated fatty acids like docosahexaenoic acid markedly impairs neuronal signal transmission but has very little effect on myelin because myelin contains almost no polyunsaturated fatty acids. Thus, the brain's autonomy regarding lipid synthesis can be betrayed by generalized malnutrition or by deficit of a specific nutrient needed to synthesize or assimilate the lipid building blocks.

The Need for Specific Dietary Nutrients

A hundred and fifty years of nutrition research has shown that to develop, mature and reproduce normally, plants and animals require a number of organic and inorganic molecules collectively known as *dietary nutrients*. These molecules cannot be made by higher organisms and without them in their food supply, a variety of essential functions, including growth and reproduction, are impaired or cease altogether.

Dietary nutrients include numerous minerals, vitamins, certain amino acids, and certain polyunsaturated fatty acids. Appropriate amounts of all these nutrients are required to achieve the normal structure and function of cells. Certain other substances that are important in metabolism or cell structure are not termed dietary nutrients because they can be made in the body irrespective of their level in the diet. These other substances include glucose and cholesterol.

Deficient intake of dietary nutrients causes reproducible symptoms, i.e. the gum deterioration and bleeding that occurs as a result of scurvy in vitamin C deficiency. These symptoms are usually characteristic and can only be corrected or prevented by adding that specific nutrient back to the diet. Such deficiencies occur in two ways: through diets low in that nutrient, or through diseases affecting the ability to absorb, utilize or retain that nutrient.

The latest report of the United Nations Children's Fund examined health and nutritional status in 80 developing countries representing about 80% of the world's population. It states unequivocally that the *'brainpower of entire nations is slipping because of a shortage of the right dietary nutrients, including, iodine, iron, vitamin A and zinc. The report claimed that inadequate iron intake reduces a child's IQ by five to seven points, while insufficient iodine reduces childhood IQ by 13 points. Iron deficiency is pervasive enough that it is estimated to reduce the gross domestic product of the most affected countries by 2%'*. The report claims that *'so ubiquitous is vitamin and mineral deficiency that it debilitates in some significant degree the energies, intellect and economic prospects of nations'*.

Supplementation with iodine or iron during infancy markedly improves growth and brain development of affected children while vitamin A supplements completely prevents a form of blindness. Though widespread, these nutrient deficiencies are still relatively specific and preventable. They are serious and debilitating but still correctable by replacement of the affected nutrients, i.e. vitamin A injection, or iodized or iron supplemented table salt. The point in relation to human brain evolution is that about a billion people are sufficiently iodine or iron deficient to cause mild to moderate impairment in mental function. This excludes the tens of millions that are severely deficient in these two nutrients.

General malnutrition as well as inadequate intake of specific dietary nutrients therefore have an adverse impact on human brain development and function that is hard to overstate. General malnutrition has devastating effects as a consequence of environmental, economic, or political disaster. In contrast, iodine and iron deficiencies are independent of disaster; their elimination only requires modest political and economic commitment.

The genetic potential of the human brain clearly cannot be developed or fully expressed under circumstances involving malnutrition or deficient intake of specific nutrients. The genetic potential of humans living in Western countries to attain their cognitive potential is only expressed today because it was learned that cretinism and goiter are caused by iodine deficiency (see Chapter 6). Prior to the 1930s, cretinism was prevalent in Europe and the USA and affected large proportions of the population, regardless of the overall affluence of the countries involved. As a result of discovering the simple link to insufficient dietary iodine, national governments now mandate iodine supplementation; if that had not happened, almost the *entire* population of the world would still be iodine deficient.

Large-scale nutrient deficiencies occur where soils are deficient but Chapters 6-8 will explain that where shore-based foods are consumed, these nutrient deficiencies are rare or do not exist at all. Because of global population growth during the past millenium, most people are no longer lucky enough to be living on the shorelines where foods naturally contain enough iodine and other essential nutrients.

If iodine deficiency prevents expression of the human brain's potential for intelligence, we have to ask how the genetic potential of the pre-human brain could ever be fully expressed in inland areas that have always provided insufficient dietary iodine relative to iodine requirements of the human brain? Even without the ice ages washing away iodine-rich soils, shore-based foods (seaweed, shellfish, fish, turtles, eggs from birds nesting near water, etc.) provide much more iodine than inland vegetation and more than meat or other animal tissues.

Iodine will be discussed separately in Chapter 6 because it is central to the case for shore-based foods as determinants of human brain evolution. My point here is that human brain evolution was dependent on changing two aspects of hominid nutrition: First, there was a general change assuring a rise in food and energy availability sufficient to permit the evolution of fatness in babies. Second, there was a specific rise in the availability of several key nutrients needed in particular by the brain. A change in neither of these aspects alone would have been sufficient; both were required to change.

In proteins, the key building blocks are the *indispensable amino acids* or, in the old terminology, the *essential amino acids*. The body has to obtain these specific amino acids from the diet or else that particular protein will not be made or will be misformed and will not work correctly. There are a couple of polyunsaturated fatty acids in phospholipids that are also indispensable. The common term for them is *essential fatty acids* because, like indispensable amino acids, they too have to be present in the diet. Docosahexaenoic acid is a good example of a polyunsaturated fatty acid indispensable for normal brain development (see Chapter 8).

How did animals (and plants) get into the situation of having an absolute dependence on certain specific minerals, vitamins, amino acids or fatty acids in the diet? Wouldn't it be advantageous to be completely independent of the composition of the diet? It would indeed be advantageous to be independent of the diet but it is more costly than it is worth. If the diet reliably contains a series of nutrients that the organism doesn't have to make, this saves the genetic programming, production of enzymes, and the metabolic costs to make those substances. Hence, it is more advantageous to be able to depend on the diet for as many nutrients as possible.

The risk is that some diets may not contain one or more of the necessary nutrients. Then the organism is out of luck and must find the appropriate diet or suffer the deficiency symptoms of lacking that nutrient. Some foods are better able to match with an organism's nutrient requirements than are others, so matching one's diet to one's nutrient needs is an important step in survival.

In a normal healthy baby whose mother is adequately nourished, breast milk supplies sufficient amounts of all nutrients for about the first six months from birth. Thus, from a strictly nutritional point-of-view, initial brain development is virtually assured for babies raised on breast milk from an adequately nourished mother.

A wide variety of foods is compatible with good nutrition. Variety is in fact the key. With the advent of agriculture, some societies have become excessively dependent on a very restricted selection of crops such as rice, cassava and corn. This is a high-risk strategy for nutritional adequacy because if the crops in question lack sufficient amounts of any of the important dietary nutrients or if the crop fails, that society will suffer the consequent deficiencies.

Two Types of Problems

The problem with important dietary nutrients comes in two broad forms: genetic errors and dietary deficiencies. The genetic errors involve an inability of the mother to absorb or use a nutrient. An example is the rare inherited disease - *phenylketonuria*, in which the indispensable amino acid, phenylalanine, cannot be metabolized properly, so it accumulates causing neurological damage. As a result, foods containing phenylalanine have to be removed from the mother's diet, or the infant will develop mental retardation.

The other damaging result of such a genetic defect is that, if there is insufficient amount of the required nutrient, a needed product (building block or enzyme) is not made in sufficient quantities. This occurs with the malabsorption of zinc in the genetic disease – *acrodermatitis enteropathica*. Genetic errors in the utilization of quite a few essential nutrients present a risk of toxicity at levels that either occur in some soils or can be induced by blockage in their metabolism.

The second broad problem with essential nutrients is the risk of their deficiency in the diet. Iodine is a classic example. Further details of the damaging effects of nutrient deficiencies on brain development will be provided in the rest of Part 1. The point in discussing them briefly here is that the brain not only has certain energy requirements but also has certain structural requirements allowing its cells to grow and communicate efficiently. Part of the structural requirement for normal brain development can be met by the body making the required molecules – proteins, phospholipids, cholesterol, etc. However, part of the structural requirement depends on sufficient dietary supply of the indispensable nutrients (vitamins, minerals, amino acids and fatty acids), which have to be present in the diet in sufficient but not excessive amounts.

One way to probe how the human brain evolved is to examine what problems arise when it isn't supplied with the nutrients it needs to develop and function normally. Regardless of whether a species is large- or small-brained, environmental variables contribute significantly to how well the brain functions and whether it will achieve its developmental potential. Broadly speaking, essential nutrients in the diet affect almost all aspects of body function and operate in all species. They place limitations on brain function mostly because they expose the reproducible and unavoidable vulnerabilities of the brain, especially during its early development.

Chapter 6

Iodine:
The Primary Brain Selective Nutrient

In a paper on the role of nutrition in human brain evolution published in 1993 and coauthored by Michael Crawford and Laurence Harbige, I introduced the term - *brain specific nutrients*. The idea was to have a collective term encompassing those nutrients needed especially for normal human brain development. This term was later revised to *brain selective nutrients* because, although they are important for the brain, these nutrients aren't specific to the brain; they are used elsewhere in the body as well. If the requirement for any one of the brain selective nutrients is not met at the correct stage of development, permanent mental retardation results. Evolving a larger brain implicitly involves either a higher requirement for or more efficient use of brain selective nutrients.

Prior to substantial expansion and increased sophistication of the brain, some (perhaps many) ecological niches would have offered sufficient food and total dietary energy to sustain healthy reproduction in early Australopithecines. The dietary availability of nutrients and energy did not likely have a direct influence on evolution of bipedalism. However, amounts of brain selective nutrients in these ecological niches would almost always have been sufficiently limiting to prevent significant brain expansion. Limited availability of brain selective nutrients in woodland and grassland niches is still likely to be a major limitation preventing full expression of the genetic potential for brain expansion in non-human primates.

Getting the nutrient intake mix right was essential for human brain evolution. The three broad stages in brain expansion (Australopithecine to *H. habilis, H. habilis* to *H. erectus,* and *H. erectus* to *H. sapiens;* see Chapter 2) would not have occurred without substantial and sustained improvements in the dietary supply of brain selective nutrients. Such improvements in diet would have required many tens to perhaps hundreds of thousands of years to take effect and would still have needed the cooperation of sustained genetic change in order to evolve advanced cognition.

As different ecological scenarios for human evolution are considered, the food sources providing an increased supply of energy, minerals, indispensable amino acids, long chain polyunsaturated fatty acids and vitamins need to be kept in mind. Where did they come from? If the 'average primate diet' was not sufficient, how were these requirements met and by which hominids? Brain expansion would have been prevented by insufficient dietary supply of brain selective nutrients in almost all geographical regions. If restricted access to brain selective nutrients was one of the keys to human brain evolution, at least one ecological niche somewhere had to provide a better supply of energy and these nutrients not generally available elsewhere.

For at least twenty years now, Robert Martin (formerly of the Institute of Anthropology in Zurich, now at the Field Museum in Chicago), Bill Leonard of Northwestern University, and others have described how an improved energy supply in the hominid diet was necessary to support evolution of the human brain. A generous and sustained dietary supply of energy was necessary not only to fuel the already relatively large energy demands of the primate brain, but to support the even higher fuel demands of further brain expansion.

My view is that abundant energy supply was a necessary prerequisite but was still insufficient to support human brain evolution because additional dietary energy alone could not meet the increasing need of the larger brain for molecules that control and refine brain function. This was the role filled only by brain selective nutrients, including at least five *brain selective minerals*: iodine, iron, copper, zinc and selenium, which control key aspects of increased cognitive sophistication. Each has a unique role in brain development and function that is described in this and the following chapter. The deficiency of any one of these minerals causes specific and well-defined neurological symptoms that, individually, would have prevented human brain evolution.

These five brain selective minerals are not necessarily more important than some vitamins, especially vitamin A, which are also probably brain selective nutrients. However, compared to several vitamins, polyunsaturated fatty acids and amino acids, it is simpler to characterize the function of essential dietary minerals because there is no capacity to synthesize them in the body.

Most minerals have different oxidation states, bind to various atoms, and their absorption or excretion may be more or less tightly regulated. However, unlike for most of the relevant vitamins, fatty acids and amino acids, the biological role of trace minerals is not complicated by varying degrees of synthesis or organic transformation in the body. Hence, the case for classifying iodine, iron, zinc, copper and selenium as brain selective minerals is simpler than it is for organic nutrients.

Iodine – The Primary Brain Selective Nutrient

Iodine is a brain selective nutrient because of its essential role in thyroid hormone production, which, in turn, is needed for normal brain development. Iodine is the only nutrient for which governments legislate supplementation in the human food supply. This is a strong indicator not only of its importance for normal human health but also the exceptionally widespread extent to which, without supplementation, humans would otherwise consume inadequate amounts of iodine. Shellfish, fish and coastal plants are rich in iodine. It is difficult (but not impossible) for humans to obtain sufficient iodine by consuming a diet lacking fish or seafood. Conversely, a modest amount of fish or seafood intake per week would essentially eliminate iodine deficiency worldwide.

At least in industrialized countries, it now is taken for granted that table salt is iodized but this has only been the case for the last 80 years. Indeed, due to political pressure for free choice by citizens, some European countries still require that *non-iodized* table salt be sold along side iodized salt. Were it not for iodination of table salt since about 1920, iodine deficiency would remain widespread today in North America and

Western Europe because, despite the affluence of these regions, fish and seafood consumption is inadequate to supply iodine requirements needed to sustain normal human brain function. Iodine deficiency is widespread in inland or mountainous areas of India, Southeast Asia, South America, and Africa.

Legislating supplementation to ensure an adequate iodine supply is necessary because most humans voluntarily eat insufficient amounts of fish or shellfish, yet without iodine supplements, the socioeconomic costs of its deficiency are exceedingly high. Iodination of table salt is the commonest method of supplementation but bread is iodized in some areas while elsewhere iodized oil is given by intramuscular injection.

The present day need for dietary iodine supplementation is the basis for iodine being the *primary brain selective nutrient*. This need for iodine supplementation indicates three important limitations about the quality of food relative to where humans live (Table 6.1): First, unless they are coastal, most habitats that might otherwise be suitable for humans do not provide diets containing adequate amounts of iodine. Distance inland from the coasts is the problem because, without sophisticated transportation systems and relative affluence, people mostly eat what is available locally. The further they are from the coasts, the less people eat fish and seafood, and the higher the incidence of iodine deficiency.

Table 6.1. Why iodine is the primary brain selective nutrient.

1.	Diets containing shore-based foods (shellfish, fish or coastal plants) are the most likely to provide sufficient iodine to meet the needs of humans.
2.	Iodine deficiency disorders are entirely preventable; indeed, do not occur, in people consuming shore-based foods.
3.	The primary sequelae of iodine deficiency involves defective mental development.

Second, iodine deficiency is entirely preventable by appropriate food selection, especially shellfish and other shore-based foods. Easily accessible shellfish like cockles and mussels are particularly rich in iodine and are abundant alone shorelines of lakes, estuaries, rivers and the sea. Eggs can be a good source of iodine, while meat and nuts are moderately good.

Third, the clinical and socioeconomic sequelae of iodine deficiency disorders revolve around impaired brain development and mental retardation. These sequelae are widespread and extreme, yet preventable. In short, optimal human brain function is incompatible with the majority of habitats that are distant from shorelines.

The neurological symptoms of iodine deficiency are debilitating and permanent (but preventable). Since people regularly consuming coastal or shore-based foods do not need iodine supplements, I make the case that the pre-human hominid brain could only have evolved into the human brain if, starting before *H. habilis*, they had greater access to a shore-based food supply. Diets based on fruits, nuts and meat, i.e. 'inland' as opposed to shore-based foods, do not supply enough iodine to have supported brain expansion. Brain expansion on the human scale necessitated sustained access to shellfish and other shore-based foods that are naturally richer in iodine than inland plants and animals.

In many regions, inland (non-coastal) communities today are still at grave risk of iodine deficiency, a risk that includes loss of fertility in addition to impaired brain development. Therefore, as hominids evolved and expanded their habitats globally over at least the last two million years, the coincident increase brain size and cognitive improvement of the hominid lineage leading to all humans was incompatible with extended occupation of areas distant from shores.

Iodine Deficiency and Hypothyroidism

Iodine is necessary for thyroid hormone production. In the short term, too little thyroid hormone causes lethargy. In the long term, mental retardation is a serious risk, especially in infants and children. Too much thyroid hormone and one becomes sweaty, irritable and hyperactive. Hence, it is important that thyroid hormone levels in the blood be well controlled, particularly during early development.

Like many other hormones in the body, a 'stimulating hormone' controls thyroid hormone secretion. For the thyroid, the controlling hormone is *thyroid stimulating hormone* (TSH). Low plasma thyroid hormone stimulates the release of TSH, which stimulates the production of more thyroid hormone. As increasing thyroid hormone levels in the blood reach the pituitary, they suppress further release of TSH, which in turn shuts off the stimulus for more thyroid hormone secretion.

This process is called *negative feedback regulation* because higher levels of the stimulating hormone raise the hormone itself (TSH in this case), which then suppresses the stimulating hormone. Control of hormone secretion, whether of thyroid hormone or others, is more complex than described here but the key element is a system of negative feedback amongst a series of hormones, one affecting the other in a loop. This results in a pulse-like rising and falling cycle of thyroid hormone production throughout the day. The feedback mechanism is essential to maintain appropriate thyroid hormone levels on which normal brain development depends.

The thyroid takes up most of the iodine entering the body. A typical adult human thyroid gland weighs about 20 grams and contains about 8 milligrams of iodine, an amount sufficient to sustain normal thyroid hormone secretion for several weeks. Almost half the iodine in the thyroid is in the form of thyroid hormone. Thyroid hormone is actually a collective term for two similar hormones - *thyroxine* and *triiodothyronine*. Iodine in the thyroid that is not yet in thyroid hormone is stored as 'pre-thyroid hormone units' called *iodotyrosines*, which are bound to a large reservoir protein called *thyroglobulin*.

As with other nutrients, iodine requirement varies with growth rate, body weight, sex, the remaining diet, climate and disease. For adult humans, there is widespread agreement that the iodine requirement is 1-2 micrograms per kilogram of body weight per day. Since a typical adult weighs about 60 kilograms, this means they need 60-120 micrograms per day of iodine, a range usually rounded for convenience to 100 micrograms per day. As with most nutrients, a somewhat higher value is recommended during pregnancy and lactation. A range of iodine intakes from 100-1000 micrograms per day is usually considered to be safe, i.e. at or above the absolute minimum required but also below a toxic intake.

During a period of iodine deficiency exceeding a couple of months, the thyroid becomes depleted of its iodine and thyroid hormone reserves and thyroid hormone secretion decreases. As usual, low plasma thyroid hormone raises TSH. Initially, triiodothyronine secretion increases disproportionately compared to thyroxine because the former is more potent and because it contains three atoms of iodine instead of the four in thyroxine, thereby economizing on iodine. Over the long term, the thyroid cannot maintain thyroid hormone production when it is iodine deficient, so the constantly low plasma thyroid hormone levels mean that the normally pulsatile stimulation of the thyroid by TSH does not let up. This constant stimulation causes the thyroid to enlarge, resulting in a lumpy swollen thyroid gland that becomes visible from outside the neck and is known as a *goiter*.

During iodine deficiency, the thyroid gland becomes more efficient at iodine recovery. However, iodine cannot really be conserved because there is no mechanism to decrease its excretion in the urine. Thus, as long as dietary intake remains suboptimal, the amount of iodine that can be incorporated into thyroid hormone continues to decrease and thyroid hormone secretion continues to decrease despite sustained stimulation by TSH. If iodine deficiency is prolonged, thyroid hormone secretion becomes chronically reduced and clinical symptoms of *hypothyroidism* develop.

In adults with no previous history of iodine deficiency, hypothyroidism is characterized by tiredness, lethargy, and increased sensitivity to cold and heat. These symptoms of slowing metabolism caused by iodine deficiency result from lower thyroid hormone levels and, consequently, a reduced signal to cells to consume oxygen and produce heat. Hypothyroidism is usually more severe in women. Mild to moderate iodine deficiency is significantly more likely to cause thyroid disease and further symptoms of iodine deficiency in women who have been pregnant compared to those that have not.

Iodine Deficiency Disorders

Thyroid hormone deficiency can be caused by a genetic defect in thyroid function totally unrelated to iodine intake or it can be due directly to iodine deficiency. The historically separate description of different forms of thyroid hormone deficiency or hypothyroidism on different continents, each confounded by different environmental circumstances, resulted in different terms being used even though they represent largely the same condition. The older terms will be described briefly here but the preferred blanket terminology now for a variety of conditions involving insufficient iodine intake and/or hypothyroidism is *iodine deficiency disorders.*

Clinical hypothyroidism or *endemic cretinism* involves shrinkage of the thyroid that can eventually result in no thyroid hormone secretion at all. In infants, clinical hypothyroidism is also known as *cretinism* or *myxedematous cretinism.* It carries a high risk of mental retardation, deaf-mutism, dwarfism, delayed sexual development, protuberant lips, flat nose, curvature of the spine, and abnormal heart rhythm. In adults, this severe form of hypothyroidism causes dry, waxy skin and is also called *myxedema.* The symptoms in adults are milder than in adolescents or children, but there is still a moderate risk of mental retardation in iodine deficient adults even if they consumed sufficient iodine during childhood. These clinical effects confirm microscopic analysis of the brain showing that human cretins have both severe loss of and damage to brain cells.

Endemic cretinism occurs in regions where goiter is widespread and severe and is associated with an iodine intake of less than 20 micrograms per day. In addition to the cognitive deficit, endemic cretinism is characterized by various neuromuscular defects including spasticity, shuffling gait and peculiar posture. Short stature, infertility, and squinting are also common. Widespread neuropsychological and physical retardation in geographic regions with endemic cretinism are not only of medical concern but they are life long, clearly making cretinism a serious socioeconomic problem as well.

The other main form of clinical hypothyroidism goes under various terms including *nervous cretinism, neurological cretinism,* or *the neurological syndrome.* In this form, goiter is still common but thyroid function tests may be normal. Strangely, thyroid test results often bear no relation to the severity of the clinical symptoms, which are seen in half or more cases of congenital hypothyroidism. The symptoms include mental retardation, deaf-mutism, abnormal eye movements, speech and language defects, dwarfism, abnormal stance and gait caused by rotation of the hips and bent knees, spasticity or spasticity-rigidity, exaggerated reflexes, and impaired ability to stand.

Nervous cretinism is common in mountainous regions of all continents. It is widespread in Papua New Guinea, Latin America,

Vietnam, eastern China, Algeria and Sicily. If iodine treatment were available prior to pregnancy, cretinism would be prevented but poverty, political instability and lack of global effort prevent this treatment from being adequately implemented. Nervous cretinism is uncommon in Central Africa despite goiter being common there. Nervous cretinism typically involves goiter but, with goiter being ten times more common than cretinism, the reverse is not necessarily true. Hence, goiter can be precipitated by factors that cause secondary iodine deficiency, i.e. impaired iodine uptake into the thyroid even when there may be sufficient iodine in the diet.

During pregnancy or infancy, hypothyroidism that is not severe enough to cause the full mental retardation of cretinism still causes significant behavioural problems including poor practical reasoning, poor hand-eye coordination, impaired hearing, poor memory, low language comprehension, reduced fine motor skills, and hyperactivity. Spasticity and related problems including poor hand to eye coordination arise from impaired gross and fine motor coordination, which indicate defective development and function of the cerebellum. Hypothyroidism may well contribute to attention deficit, which also occurs in iron deficiency.

Despite the obvious importance of these symptoms, clinicians typically classify such intermediate cases as 'minimal brain dysfunction'. IQ tests may initially score within the normal range but age-adjusted scores tend to decrease. By nine years old, about a fifth of children born hypothyroid are in special education. The origin of these problems is usually *fetal* hypothyroidism because the degree of developmental delay during childhood correlates directly with the extent of hypothyroidism that is already present at birth. Without exception, these developmental problems are difficult to correct.

Serious hearing impairment occurs in hypothyroidism, probably because of abnormal ossification of the *Organ of Corti* in the inner ear. Even when hearing is not impaired, learning from the spoken word and verbally expressing ideas are impaired, indicating that the assimilation of auditory information is impaired. It is interesting to note that hearing is an aspect of brain function that has been independently shown to have a particularly high energy requirement. Expansion of the auditory region in humans undoubtedly facilitated the evolution of speech uniquely in humans. Hence, two key aspects of human evolution (speech and brain enlargement) had an increasing requirement for energy and are also vulnerable to inadequate iodine status.

Goiterogens

Goiterogens or goiter-causing substances in food are a major contributor to iodine deficiency disorders and complicate the etiology of different forms of cretinism in several regions of the world. Cassava is probably the most important goiterogenic food both because of its relatively high content of two iodine binding substances, *hydrocyanic acid* and *linamarin*, and because of its widespread consumption as a staple food of humans. The body detoxifies hydrocyanic acid to *thiocyanate*. However, thiocyanate is still a strong goiterogen that is excreted in larger quantities, thereby depleting the body of iodine.

Other goiterogenic foods also consumed widely by humans include maize, cabbage, rape, soybean and mustard. These foods are frequently used in feed for farm animals where they are also goiterogenic. They reduce the iodine content of animal tissues, which reduces iodine available to humans consuming meat or dairy products from these animals. Though a less common cause of iodine deficiency, peanuts and pine nuts also contain mild goiterogens.

Because they bind to dietary iodine and impair the body's ability to absorb it, goiterogens can induce iodine deficiency despite an apparently adequate level of iodine in the diet. Hence, regardless of iodine intake, TSH is the best measure of iodine that is functionally available for the thyroid.

Iodine Deficiency in Developed Countries

During the ice ages, continental and alpine glaciation extensively eroded soils of the Northern Hemisphere. Adding to the soil-depleting effects of glaciation, extensive flooding and leaching has made soils in most temperate and tropical areas of the upper Northern Hemisphere more deficient in iodine than other minerals. This loss of iodine from the soils is especially evident in alpine and mountainous areas, whether temperate or tropical, so it is in these regions where human iodine deficiency is still commonly found.

Even with mandatory iodation, iodine deficiency is still sufficiently widespread and deleterious to be currently ranked the world's most serious nutritional deficiency, adversely affecting nearly a billion people or more than a fifth of the world's population. It is staggering to think what the extent of iodine deficiency would have been if food or salt iodation programs had not been implemented. Therefore, the sheer scale of this problem emphasizes how few people are currently occupying areas that naturally provide adequate dietary iodine.

Iodine deficiency disorders have a significant socio-economic impact in developing countries, especially those that are poor and have populations subsisting on vegetarian diets that contain goiterogens. However, iodine deficiency is not just a problem in poor countries.

Less than a decade ago, the World Health Organization reported that up to 10% of Europeans, especially women, are presently at risk of iodine deficiency. Other reports surveying large numbers of people, including the National Health and Nutrition Examination Survey (NHANES III) in the USA, suggest that mild hypothyroidism presently goes undetected in upwards of 10 million Americans, and upwards of 40 million Europeans. On the basis of urinary iodine excretion decreasing by about 50% over the past 20 years, recent reports suggest that over 20% of Australian women are also at risk of being mildly to moderately iodine deficient.

Measuring Iodine Status

Plasma TSH is the best test of thyroid function because it rises dramatically in response to a modest decline in thyroid hormone. There is no precise point at which declining thyroid function becomes hypothyroidism. In practice, a maximum blood value for TSH has been chosen, above which intervention with thyroid hormone supplements is started. That TSH value is usually 4 milliunits per liter (mU/L) of plasma. This compares to a maximum 1.5 mU/L in those considered to be totally free of the risk of hypothyroidism. In some parts of the USA, up to 10% of adults have a TSH above 5 mU/L so, by definition, they are clinically mildly hypothyroid.

Concern is emerging that hypothyroidism is significantly under-diagnosed because the upper limit of normal plasma TSH has been set too high, i.e. that mild thyroid dysfunction actually starts with a TSH lower than 4 mU/L. Some British experts feel that hypothyroidism really starts with TSH above 2 mU/L. This concern is well supported by studies in different countries involving millions of people. Cost-benefit analysis suggests screening for hypothyroidism is as effective as screening for hypertension in the same age group. Mild hypothyroidism is associated with diabetes, hypercholesterolemia, osteoporosis, hyperlipidemia, hypercholesterolemia, hyperhomocysteinemia, and cardiovascular and neuropsychiatric disease, especially in the elderly.

The persistent problem of mild hypothyroidism and mild to moderate iodine deficiency in affluent countries seems to be independent of moderately high supplementation of table salt with iodine. This is because although salt intake is necessary to prevent hypothyroidism in most people, salt intake has been decreasing for many years due to the perceived risk of excess dietary salt for hypertension. Iodine intake is also decreasing not only because of insufficient intake of foods that are the best sources of iodine, i.e. fish and shellfish, but also because of lower intake of foods that are moderately good sources of iodine, i.e. milk, eggs and meat.

Though richer in iodine than fruits or vegetables, many common foods that are moderately good sources of iodine are often shunned in affluent regions because they are perceived to be unhealthy. Indeed, the irony is that an apparently healthy diet consisting of cereals, vegetables and fruit is very likely to be iodine deficient. This risk of iodine deficiency occurs in part from the low iodine content of these foods and in part from the higher intake of goiterogens in popular vegetarian food ingredients like soy. This problem can only be avoided by reducing intake of goiterogen-containing foods while increasing consumption of milk, eggs, fish and/or shellfish.

Iodine Content of Foods

For at least a thousand years, various coastal populations have known that eating seaweed is a simple and effective way to treat goiter. About 200 years ago, iodine itself first became directly associated with the remedial effects of seaweed and shore-based foods. The content of iodine in the major categories of foods available today is shown in Table 6.2.

Present day dietary sources of iodine vary with the degree of iodination during food processing. Modern day food processing practices are largely irrelevant to nutrient availability prior to one hundred years ago so the notable presence of iodine in bread and other baked products is not relevant to human brain evolution. Thus processed foods are not accounted for in Table 6.2. However, modern day iodine supplementation of processed foods is relevant to minimizing the occurrence of hypothyroidism that would otherwise be widespread.

Table 6.2. Iodine content of various major food groups.

Food Group[1]	Iodine Concentration[2] (micrograms per 100 grams)	Amount Needed [3] (grams per day)
Shellfish	10-28	680 (360-1000)
	(150 in cockles, mussels) [4]	(67) [5]
Eggs	53	190
Vegetables	1-3	4,200 (3,300-5,000)
	(20,000 in dried seaweed)	(<1)
Fruit	1-5	6,000 (2,000-10,000)
	(9 in strawberries)	(1110)
Nuts	5-10	1,500 (1,000-2,000)
	(20 in peanuts)	(500)
Cereals	6	1,670
Pulses	2	5,000
Milk	15	6,670
Meat	5-10	1,500 (1,000-2,000)
Fish	50-110	150 (90-200)
	(250 in haddock)	(40)

[1]These groups of foods are arbitrary but encompass all unprocessed foods.
[2]Compiled from a variety of reference sources. Where available, a range of values is given in brackets.
[3]Amount of the food category needed to satisfy the daily iodine requirement, which has been averaged to a single, commonly accepted value of 100 micrograms per day for adult, non-pregnant humans.
[4]The amount of iodine in the foods richest in iodine (even if rare).
[5]Best case scenario, using the richest dietary sources of iodine.

Cow's milk, fruit and vegetables are uniformly low in iodine, so much higher volumes of these foods are required to meet the human dietary requirement for iodine (Table 6.3). Second, the iodine content of cereals and pulses (beans, peas and legumes) is somewhat higher than in fruit or vegetables. Third, fish, meat and nuts are moderately rich in iodine but still lower than shellfish or eggs. Thus, any intake of eggs or shellfish dramatically decreases the amount of any other types of food that would be needed to meet iodine requirement.

Table 6.3. Amount (grams) of each food group required to meet the daily iodine requirement in non-pregnant, non-lactating adult humans. Variation in abundance of these foods, e.g. with season or need for domestication, and any restrictions on bioavailability of iodine from these foods, has not been taken into consideration in calculated quantities required.

Fish	150
Eggs	190
Shellfish	210
Meat	1,500
Nuts	1,500
Cereals	3,200
Pulses	3,700
Vegetables	4,200
Fruit	6,000
Milk	6,670

Iodine and the Brain

Little is known about precisely how iodine deficiency impairs brain development. Iodine and thyroid hormone are involved in energy metabolism and since the brain has a high metabolic rate and demand for energy it seems logical that this vulnerability could lead to impaired brain development and function in iodine deficiency. Several steps in ketogenesis are inhibited by nearly 50% in iodine deficient rats. Since iodine deficient young animals have an impaired ability to generate ketone bodies, this implicates iodine in the well-established role of ketones in brain energy metabolism.

Delayed neurological development in iodine deficiency includes impaired production of myelin, a process also linked to availability of the ketone bodies, beta-hydroxybutyrate and acetoacetate, which are the primary substrates for brain lipid synthesis during early development. Thus, iodine deficiency impairs at least two key aspects of lipid metabolism in the developing mammalian brain - ketogenesis for energy metabolism and myelin synthesis. It is unknown whether iodine has other roles in brain energy metabolism independent of its effects on brain lipid metabolism.

Iodine and Human Evolution

Humans are unable to conserve iodine when they become iodine deficient. Rather, in iodine deficiency, urinary excretion of iodine continues as usual. The inability of humans to conserve iodine seems to have arisen and persisted because, early in human evolution, diets were rich in iodine and there was no need to conserve iodine. If hominids evolved on shorelines, iodine would have been abundant in the diet. Hence, controls did not evolve to reduce its excretion because diets from at least the time of *H. erectus* until the advent of agriculture were not limiting in iodine. The only place where dietary iodine is abundant and, hence, where this oversight could arise without adversely affecting brain expansion is on the shorelines.

No other species in the wild, including primates, exhibits symptoms of iodine deficiency while consuming their routine diet. The implication is that if symptoms of deficit are normally uncommon, then intake of a nutrient is probably not limiting for reproduction, growth or development. Since it is unreasonable to think that the human requirement for iodine (or other nutrients) dramatically increased in the past few thousand years, this unfortunate situation seems to have arisen from increasing human occupation of geographical regions ill-suited to human iodine requirements, i.e. inland or mountainous regions distant from shorelines and shore-based foods.

The modern day prevalence of iodine deficiency disorders and the exquisite sensitivity of the developing brain to insufficient iodine or thyroid hormone are perhaps the most important evidence for the shore-based hypothesis of human evolution. How could the brain of early *Homo* have expanded and become increasingly sophisticated in the face of this major endocrine and dietary hurdle?

Insufficient intake of brain selective nutrients like iodine prevents the brain of *H. sapiens* from developing or functioning normally in modern times so, regardless of genetic potential, the ongoing presence of such a nutrient limitation in the diets of the Australopithecines would have prevented substantial expansion of cognitive capacity. Insufficient dietary iodine would have prevented emergence of the larger, more capable brain of *H. erectus*.

Once pre-human hominids found themselves in an environment providing more iodine and with the brain slowly expanding, vulnerability to inadequate iodine intake would then be more acute in the larger-brained individuals. In effect, the larger pre-human brain became even more dependent on finding an appropriate diet to sustain the higher iodine intake needed for its normal development. Hominids choosing to remain in a lower iodine environment, eg. the *Pananthropus* genera, would have foregone brain expansion but would not be otherwise compromised. Those entering a truly low iodine environment would have become sterile.

If the human predilection to consume iodine deficient diets and become mildly to moderately hypothyroid does not change, this oversight may well constitute a new environmental stimulus to some future degenerative change in the cognitive capabilities of *H. sapiens* in iodine-deficient regions.

Chapter 7

Iron, Copper, Zinc and Selenium: The Other Brain Selective Minerals

Iron

Despite knowledge of the health attributes of iron since ancient times, and its abundance in the earth's soils, iron deficiency has become a widespread, serious nutritional problem in humans. It is much more of a problem in developing compared to industrialized countries but is still prevalent wherever the population is poor. Chronic mild to moderate iron deficiency is common because it can remain undetected for months in infants, children and adults. In many areas of the world it represents a persistent, significant obstacle to normal neurological development.

Iron is required for the normal production and function of several important proteins, notably *hemoglobin* for oxygen transport in blood, and *myoglobin* for oxygen transport in muscle cells. Hemoglobin contains the main pool of iron in the body. Red blood cells carry hemoglobin so a low red cell level in blood (*anemia*) reveals low iron status. As plasma iron starts to decline at the onset of iron deficiency, additional iron is released from body stores in the form of two proteins – *ferritin* and *hemosiderin*. If this additional iron does not sustain or restore plasma iron, yet another iron-rich plasma transport protein – *transferrin* – will release iron.

If iron deficiency persists, hemoglobin synthesis will decrease which, in turn, reduces production of red blood cells. In mild *iron deficiency anemia,* red cell iron levels may be normal but more subtle markers of iron deficiency usually are present, i.e. low serum ferritin. In severe iron deficiency, this results in *hypochromic microcytic anemia* or low amounts of red blood cells each low in hemoglobin.

Myoglobin in muscle and hemoglobin in red blood cells are similar proteins because both have a structural core called *heme*, which is the part of the protein that actually traps the iron. Dietary iron consumed as meat or blood products is therefore known as *heme iron*, whereas iron in other dietary constituents or foods is *non-heme iron*. Unlike in many plant foods, iron in meat is easily digested and absorbed, i.e. it is readily *bioavailable*. Despite the value of meat as a source of iron, low to negligible meat intake in many areas of the world makes meat a minor source of iron for millions of people, especially in poorer countries.

The ability of muscles to obtain oxygen determines their work capacity so impaired ability to make hemoglobin due to iron deficiency is a major cause of poor strength and early onset of fatigue. Mild to moderate anemia is associated with lower oxygen uptake from the lungs and is strongly positively correlated with poor performance in exercise tests. Weakness and fatigue are commonly complicated by a tendency to be less resistant to infections in iron deficient individuals. Chronic iron deficiency also occurs due to menstrual blood losses or gastrointestinal bleeding caused by parasitic infection in the gut.

Iron is needed for 'activation' of oxygen by several *oxidase* and *oxygenase* enzymes, and in the *cytochrome* proteins used to transport electrons during fuel production in mitochondria. Iron's role in these enzymes derives from an interplay between its two *oxidation states*, *ferrous* (Fe^{+2}) and *ferric* (Fe^{+3}), on which several steps in cellular respiration and metabolism depend. Iron deficiency impairs control of body temperature by interfering in conversion of thyroxine to triiodothyronine. This reduces the ability to conserve heat and contributes to coldness in the extremities as well weakness and fatigability in iron deficiency. Like iodine, sufficient iron is essential for the body to effectively use oxygen and metabolic fuels and to meet its energy requirements.

Iron absorption is highly variable depending on the food source in question. Its absorption is facilitated by other nutrients such as vitamin C. Like with zinc, however, iron absorption is inhibited by *phytate*, a molecule abundant in plants rich in fiber, including cereals and legumes (beans and pulses). Phytate binds to trace minerals thereby inhibiting their absorption. This is a well-established problem that affected sedentary paleoagricultural tribes such as the Pima of the southwestern USA, who at various times since the advent of agriculture have been severely anemic and iron deficient due to high intake of corn, which is rich in phytate. The problem of poor absorption of iron from plant-based foods can be ameliorated by the presence in the diet of animal foods, i.e. fish or meat, especially liver, because they are richer in iron and do not contain phytate (see Table 7.1).

Table 7.1. Iron content of the main food groups.

Food Group[1]	Concentration[2] (milligrams/100 grams food)	Amount of Needed [3] (grams/day)
Shellfish	1-3	800 (400-1200)
	15 in winkles, 25-40 in cockles[4]	30[5]
Eggs	2	600
Vegetables	0.4-1.0	2,100 (1,200-3,000)
	5-7 in parsley, endives	200
Fruit	0.2-0.8	3,700 (1,500-6,000)
	1.0 in berries, olives; 1.5 in avocados	1,000
Nuts	1-3	800 (400-1,200)
	4 in almonds	300
Cereals	0.2-4	3,100 (300-6,000)
Pulses	2-8	370 (150-600)
Meat	1-3	800 (400-1,200)
	3-7 in organ meats; 10 in liver	120
Fish	0.3-0.4	3,500 (3,000-4,000)
	1.5 in mackerel	800
Milk	0.05	24,000

[1]These common food groups are arbitrary but encompass all unprocessed foods.
[2]These data are compiled from a variety reference sources.
[3]Average amount of the food category needed to satisfy the daily iron requirement (12 milligrams per day). Where available, a range is given in brackets.
[4]The amount of iron in the foods richest in iron (even if rare).
[5]Best case scenario, using the richest dietary sources of iron.

A major global cause of long-term iron deficiency is poor absorption of iron from cereal-based diets. In poor areas, *geophagia* or *pica* (eating earth, especially clay) contributes to iron deficiency because soils contain compounds binding to iron and preventing its absorption. However, if adults initially have adequate body iron stores, short term decreases in dietary iron intake rarely cause acute symptoms of iron deficiency because, unlike for iodine, iron excretion is well controlled and can be reduced to prevent further losses.

Currently, 20% of women in the USA are iron deficient enough to have no detectable ferritin. Half of these women are anemic, i.e. have low hemoglobin or red cell count in the blood. Complicating the detection and treatment of iron deficiency is the fact that chronic inflammatory diseases commonly raise serum ferritin. This effect therefore masks the effect of iron deficiency and can misleadingly suggest that some people with chronic inflammatory disorders such as arthritis have normal iron status when in fact they are iron depleted.

Iron and Brain Function

Historically, the relationship of iron deficiency to impaired brain function originated in empirical observations concerned with fatigue, weakness and headache amongst patients subsequently shown to be anemic. Brain iron concentration and neurotransmitter levels are reduced by dietary iron deficiency. Betsy Lozoff (University of Michigan) has pointed out that vulnerability to iron deficiency is critical during early development because that is when the brain growth is most rapid and because infants are also more likely to become iron deficient than are adults.

Iron deficiency appears to be a principal cause of neurodevelopmental delay and emotional fragility seen in malnourished children. Ernesto Pollitt (University of California at Davis) is one of the world's leading experts on iron deficiency in humans. He has shown that iron deficiency has a significant and specific effect on visual attention, concept acquisition, verbal scores, and school achievement. Some of the decline in cognitive performance in iron deficient children appears to be coordinated via changes in brain receptors for two neurotransmitters, *dopamine* and *gamma-aminobutyric acid*. Elevated plasma glucose and insulin during iron deficiency and impaired transport into the brain of the amino acid, *valine,* may also be relevant.

Lower oxygen uptake in the brain caused by insufficient iron appears to contribute to poorer performance on cognitive tests in iron deficient children. However, performance on tests of cognitive function is modified by arousal, attention and, especially, motivation. The emotional reaction to novelty *(affect)* plays an important role in behavioural testing. Abnormal affect includes hesitancy, fearfulness, and tenseness. It can result in unreactivity to or disinterest in test materials. Familiarity with the examiner is also a factor in affect. Pollit's studies make it clear that getting to the root of the role of iron in neurodevelopment is confounded by deterioration in affect in iron deficient children.

Copper

Normal myelin synthesis requires copper. *Hypomyelination* occurs during dietary copper deficiency, and this contributes to poor control of the skeletal muscles because the nerves do not conduct signals faithfully unless they are fully insulated by myelin. The link between myelination and copper was first observed in copper deficient lambs, in which hypomyelination causes a shaky gait called *swayback*. Swayback results from either insufficient copper in pastures or a combination of low copper intake plus a dietary excess of other minerals that competitively reduce copper absorption.

In humans, dietary copper deficiency is rarely if ever severe enough to cause the equivalent of swayback. However, *Menke's disease* is a genetic abnormality in copper absorption causing severe copper deficiency, hypomyelination and mental retardation. Copper deficiency in Menke's disease occurs because copper absorbed from the diet is not released from intestinal cells to the rest of the body. The symptoms include progressive mental retardation, hypothermia, and seizures. Death during infancy is common.

In addition to normal myelination requiring sufficient copper, at least three enzymes involved in neurotransmitter synthesis and degradation also require copper as a cofactor: They are *tyrosinase,* which converts the amino acid, *tyrosine*, to the neurotransmitter *dopamine*, and also converts dopamine to *dopaquinone*; *dopamine-beta-hydroxylase*, which converts dopamine to another neurotransmitter - *norepinephrine*, and *monoamine oxidase*, which inactivates monoamine neurotransmitters, including noradrenaline, serotonin, and dopamine.

Connective tissue defects affecting bone and arterial structure also occur in Menke's disease because copper is needed for synthesis of two proteins that are part of the connective tissue of bone, cartilage and arteries (*collagen* and *elastin*). Deficient copper intake therefore results in weak bone and easily ruptured arteries. The copper-dependent enzyme used to cross-link or strengthen collagen and elastin is *lysyl oxidase*.

Copper deficiency raises blood cholesterol and impairs glucose metabolism, thereby mimicking the effect of atherosclerosis and diabetes, respectively. Mild to moderate copper deficiency may well exacerbate the risk of heart disease in middle-aged individuals in industrialized countries. Several proteins with antioxidant properties contain copper, including *ceruloplasmin* as well as two different forms of the *superoxide dismutase* enzymes.

Copper also has several important links to iron and energy metabolism. First, in addition to its antioxidant activity, the copper transport protein, *ceruloplasmin*, is needed for iron incorporation into hemoglobin. Therefore, regardless of iron status, a common symptom of copper deficiency is anemia. Second, like iron, copper is also required for normal function of at least one of the cytochromes in the electron transport chain *(cytochrome C oxidase)*, and is therefore needed to produce ATP for cell metabolism. Cytochrome C oxidase has high activity in the brain, when cellular energy requirements are very high.

Hence, copper is a brain selective mineral for two broad reasons: First, it has several essential biological roles relevant to brain function, i.e. in neurotransmitter and myelin synthesis, in elastin used for blood vessel structure, in antioxidant protection, and finally, with iron, for cellular energy metabolism. Expanding brain size and complexity of connections between neurons increased the need for the supporting processes, several of which depend on copper, including antioxidant protection of the additional membrane lipids which are easily damaged by free radicals and lipid peroxidation, additional capacity for cellular energy production, more myelination for integrity of many more signals, and additional blood flow to meet the increasing energy requirement of the brain.

Second, many human diets around the world are marginally sufficient in copper. Shore-based foods, nuts, and organ meats, especially liver, are the best sources (Table 7.2). Copper excretion is tightly controlled, which helps protect against deficiency symptoms despite low copper intake in most countries. Copper content of many human diets is marginal, a situation aggravated by the fact that copper absorption is inhibited by excess vitamin C, iron, molybdenum or zinc.

Table 7.2. Copper content of the main food groups.

Food Group[1]	Concentration[2] (milligram/100 grams)	Amount Needed [3] (grams/day)
Shellfish	0.2-0.4	900 (625-1,250)
	0.8 in shrimp; 7 in oysters, whelks[4]	35[5]
Eggs	0.1	2,500
Vegetables	0.06-0.2	2,700 (1,250-4,200)
	0.6 in mushrooms	420
Fruit	0.03-0.2	4,800 (1,250-8,300)
	0.4 in pineapples	625
Nuts	0.2-0.5	900 (500-1,250)
	1.0 in Brazil nuts	250
Cereals	0.03-0.2	4,800 (1,250-8,300)
Pulses	0.5-2 (beans, lentils, chick peas)	300 (125-500)
Meat	0.1-0.3	1,700 (833-2,500)
	2-10 in liver	25
Fish	0.05-0.2	3,100 (1,250-5,000)
Milk	0.02	12,500

[1]These food groups are arbitrary but encompass all unprocessed foods.
[2]These data are complied from a variety reference sources.
[3]Average amount of the food category needed to satisfy the daily copper requirement (2.5 milligrams per day).
[4]The amount of copper in the foods richest in copper (even if rare).
[5]Best case scenario, using the richest dietary sources of copper.

Zinc

Zinc is needed for the normal activity of over one hundred enzymes, especially those involved in digestion of food, and in the synthesis of DNA and proteins. It is a structural component of several proteins and enzymes involved in gene expression. These roles make it essential for normal growth, development, sexual maturity, reproduction, immune function, taste and appetite, and tissue repair. Cell division requires DNA replication so cells that must be frequently replaced are most vulnerable to zinc deficiency, i.e. those in the skin, immune system, and in the lining the intestine.

The requirement for zinc is linked to nitrogen intake, which is primarily used for protein and tissue synthesis; if protein intake is low, the limiting effect of zinc on these processes is less obvious because growth is already inhibited by low nitrogen intake. The body does not store zinc so regulation of body zinc levels is dependent on the efficiency of zinc absorption from the intestine.

Meat is a good source of zinc but shellfish, especially oysters, is the best dietary source of zinc. Phytate is a significant component of cereals that inhibits mineral absorption, so dietary insufficiency of zinc occurs in areas in which the diet is rich in cereals. Thus, zinc is less available from cereals than would appear to be the case from their zinc content alone (Table 7.3). Pica or clay consumption in extremely poor populations causes *hypogonadism* and dwarfism due in part to chronic zinc deficiency. High intakes of copper, calcium and iron are competitive with zinc and can also cause its deficiency. In addition, areas where zinc intake is marginal are often plagued by low food availability and chronic infectious disease, which exacerbates zinc deficiency.

Zinc deficiency can also be secondary to genetic disorders affecting its absorption, or to diseases of malabsorption or liver disease. *Acrodermatitis enteropathica* is a rare genetic disease in which zinc absorption is impaired. It causes skin lesions, gastrointestinal malabsorption, growth failure, and hair loss.

Table 7.3. Zinc content of the main food groups.

Food Group[1]	Concentration[2] (milligrams/100 grams)	Amount Needed[3] (grams/day)
Shellfish	2-5	500 (280-700)
	7 in whelks, 10-100 in oysters	14
Eggs	1.5	930
Vegetables	0.1-0.4	8,700 (3,500-14,000)
	0.9 in parsley, 1-2 in green beans	700
Fruit	0.1-0.3	9,300 (4,700-14,000)
Nuts	2-4	500 (350-700)
Cereals	0.6-1.0	1,900 (1,400-2,300)
Pulses	3	470
Meat	1-4	900 (350-1,400)
	7 in calf's liver	200
Fish	0.3-2.0	2,700 (700-4,700)
	3 in herring; 6 in walleye pike	230
Milk	0.03	47,000

[1]These food groups are arbitrary but encompass all unprocessed foods.
[2]These data are compiled from a variety reference sources.
[3]Average amount of the food category needed to satisfy the daily zinc requirement (14 milligrams per day).
[4]The amount of zinc in the foods richest in zinc (even if rare).
[5]Best case scenario, using the richest dietary sources of zinc.

Zinc is a brain selective mineral because of the rapid onset of zinc deficiency in infants and children, and because zinc is highly concentrated in *mossy fibers* of the hippocampus where it appears to be involved in *enkephalin* binding to opiate receptors. Zinc is also needed for normal metabolism of one of the neurotransmitters – norepinephrine. Learning and memory are impaired in young, zinc deficient animals, partly because zinc is also needed in the metabolism of polyunsaturated fatty acids, notably in the synthesis of *arachidonic acid*.

Selenium

Selenium is best known for its role in the normal activity of a family of antioxidant proteins called *glutathione peroxidases*, which are used for defense against the damage produced when oxygen attacks unsaturated lipids, releasing free radicals and causing *lipid peroxidation,* or rancidity. Polyunsaturated fatty acids such as arachidonic acid and docosahexaenoic acid are very susceptible to lipid peroxidation because of their multiple double bonds (four and six, respectively). Glutathione peroxidase has a key role in protecting docosahexaenoic acid in the brain and eye during early development. Vitamin E is also involved in antioxidant protection, and selenium deficiency is exacerbated by vitamin E deficiency. Another selenium containing protein, *thioredoxin reductase,* is an antioxidant but is also involved in DNA synthesis.

Free radicals can be damaging but they are also needed in the immune response so their appropriate production yet confinement is essential. Immune function is impaired by selenium deficiency, thereby increasing potency of viruses that normally are not very harmful. This increased vulnerability to infection may affect susceptibility to heart disease and chronic inflammatory diseases. Deficient intake of selenium is also involved in *Keshan-Beck disease,* which causes cardiomyopathy in children and is widespread in rural China.

Selenium is involved in several processes that involve two other brain selective minerals – iron and iodine. For instance, heme production is not only dependent on iron but is also selenium-dependent, making selenium important in the capacity of tissues, especially the brain, to obtain oxygen and ramp up energy metabolism. Selenium is essential for the activity of the enzyme, *iodothyronine deiodinase*, which is required to remove one iodine atom from *thyroxine*, thereby converting it to *triiodothyronine* in target tissues outside the thyroid itself. Depending on the tissue, three different versions of this enzyme exist but all three are selenium-dependent.

Plasma thyroid hormone levels are not necessarily affected in selenium deficiency but the efficacy of thyroid hormone action is impaired by selenium deficiency because triiodothyronine production is impaired. In effect, selenium deficiency contributes significantly to iodine deficiency disorders. In addition to its role in reducing lipid peroxidation, the key role of selenium in iodine metabolism is perhaps the most important reason selenium is a brain selective nutrient.

Table 7.4. Selenium content of the main food groups.

Food Group[1]	Concentration[2] (micrograms/100 grams)	Amount Needed [3] (grams/day)
Shellfish	30-40	300 (250-330)
	60-70 in oysters[4]	150[5]
Nuts	1-7	5,500 (1,430-10,000)
	200 in Brazil nuts	50
Meat	1-14	5,000 (700-10,000)
	200 in kidney	50
Eggs	11	900
Fish	10-30	660 (330-1,000)
	60 in herring	170
Cereals	3-9	2,200 (1,100-3,300)
	53	190
Pulses	2-10	3,000 (1,000-5,000)
	100 in lentils	100
Vegetables	1-3	6,700 (3,300-10,000)
	12 in mushrooms	830
Fruit	1-5	6,000 (2,000-10,000)
Milk	1-10	5,500 (1,000-10,000)

[1]These food groups are arbitrary but encompass all unprocessed foods.
[2]These data are compiled from a variety reference sources.
[3]Average amount of the food category needed to satisfy the daily selenium requirement (100 micrograms/day).
[4]The amount of selenium in the foods richest in selenium (even if rare).
[5]Best case scenario, using the richest dietary sources of selenium.

Dietary selenium generally occurs as *selenomethionine* or *selenocysteine,* which are associated with proteins, so it is generally low in fruits and vegetables because they are low in protein (Table 7.4). Meat is a moderately good source of selenium. Fish, shellfish and other shore-based foods are the best sources of selenium. The selenium content of plants depends on the selenium content of the soil, which can vary from deficient in one region to toxic in another. In several areas of the world, including the far western USA, several areas in China and all of New Zealand, selenium levels in the soil are insufficient to feed to livestock without selenium supplements. In contrast, some plants in the central USA accumulate sufficient selenium to be toxic to animals.

Brain Selective Minerals in the Diet

Knowing the human daily requirement for these five brain selective minerals and their concentration in various foods allows the calculation of how much of each food group would be needed if that food group alone supplied the daily requirement for these minerals. Table 7.5 shows that 900 grams of shellfish would be required to supply the daily needs for all five minerals. This is considerably less than for any other food group. Except for their low copper content, eggs would be as good a source of these minerals as seafood. Fish are the best source of iodine and selenium but are poor in copper, zinc and iron so they rank below seafood and eggs. Pulses supply more copper, iron and zinc than the equivalent weight of seafood but pulses have much less iodine and selenium than seafood, so they rank further down the list.

Table 7.6 simplifies the numerical values in Table 7.5. Note that this ranking is based on the amount the least to most abundant of these five minerals in each food category. Note also that the amounts of food required seem high (even 900 grams of shellfish) but are based on food composition tables and established nutrient requirements in humans. Requirement values aim to avoid deficiency in at least 97% of the population, so they may exceed the requirement of many individuals. Hence, the intakes shown in Tables 7.5 and 7.6 reflect this stipulation.

Table 7.6 simply reconfirms in a different format that shellfish is best able to meet the adult requirement for these brain selective minerals. Thus shellfish is relatively poor in copper but rich in iodine and iron so it takes more shellfish to meet the daily requirement for copper than would be needed to meet the iron or iodine requirement. On the other hand, pulses are rich in copper but much higher amounts of pulses are needed to meet the requirement for iodine than for copper. Eggs or nuts are the second best sources of most of these minerals. Copper is the least abundant brain selective mineral in three food groups (shellfish, eggs and cereals), zinc is the least abundant in three food groups (vegetables, fruit and milk), selenium is least abundant in two food groups (meat and nuts), while iodine is least abundant in one food group (pulses; Table 7.6).

Table 7.5. Amount of each major food group required to meet the daily requirement for five brain selective minerals.

Food Group	Iodine	Iron	Copper	Zinc	Selenium
Shellfish	680[1]	800	(900)	500	300
Eggs	190	600	(2,500)	930	900
Fish	150	(3,500)	3,100	2,700	660
Pulses	(3,700)	370	300	470	3,000
Cereals	3,200	3,100	(4,800)	1,900	2,200
Meat	1,500	800	1,700	900	(5,000)
Nuts	1,500	800	900	500	(5,500)
Vegetables	4,200	2,100	2,700	(8,700)	6,700
Fruit	6,000	3,700	4,800	(9,300)	6,000
Milk	6,670	24,000	12,500	(47,000)	5,500

[1]Values are for non-pregnant, non-lactating adult humans and are shown in grams. The nutrient shown in brackets is the most limiting in each food group, i.e. it is the mineral for which the largest amount of each food group would be required daily.

Table 7.6. Rank order of the major food groups according to their ability to meet the needs of five brain selective minerals.

Food Group (minimum amount required/day)	Ranking of the least to most abundant mineral in each food group
Shellfish (900 grams)	Copper >> selenium > zinc > iron > iodine
Eggs (2,500 grams)	Copper >> zinc = selenium > iron > iodine
Fish (3,500 grams)	Iron > copper > zinc > iodine > selenium
Pulses (3,700 grams)	Iodine > selenium >>> zinc > iron > copper
Cereals (4,800 grams)	Copper > iron > selenium > zinc > iodine
Meat (5,000 grams)	Selenium >> copper > iodine > zinc > iron
Nuts (5,500 grams)	Selenium >> iodine > copper = iron > zinc
Vegetables (9,000 grams)	Zinc > selenium > iodine > copper > iron
Fruit (9,000 grams)	Zinc > selenium = iodine > copper > iron
Milk (47,000 grams)	Zinc >> iron >> copper >> iodine > selenium

This ranking is based on the average amount of each food group (grams/day) that is needed to meet the daily needs of the least abundant of these five minerals in each food group. It is based on the concentration of these minerals in a range of foods in (Table 6.2 and Tables 7.1 to 7.4) as well as on the minimum daily requirement for each of these minerals.

Based only on the concentration of these five minerals in these ten food groups, lower amounts of shellfish than any other food group (less than a kilogram per day) would be needed to meet the daily needs of humans for these brain selective minerals. The next best source is eggs but about 2.5 kilograms would have been required. Pulses and fish are next at about 3.5 kilograms each, followed by cereals, meat or nuts at about five kilograms each. No one food group totally lacks all of these minerals but milk is clearly the poorest source by a wide margin.

Shellfish is the best all round source of these brain selective minerals. Thus, inclusion of any amount of shellfish in the diet would have helped improve supply of these minerals to the human (or pre-human hominid) brain. Because shellfish is such a good source of brain selective minerals, and because major inland areas of most continents lack sufficient dietary sources of iodine, selenium and iron, it seems that intake of shellfish can't help but contribute to normal human brain function.

These rankings are not corrected for *bioavailability* (mostly absorbability) of these minerals. Nutrient bioavailability is not compromised by shellfish, eggs, fish or meat, but several minerals are markedly less available from plants, especially those containing phytate. These rankings are also not corrected for accessibility of the food groups in paleolithic times. Domestication in the past 10,000 years markedly increased availability of pulses, cereals, meat and milk, but had not occurred a million years ago. In many cases, especially for species listed in these tables, fishing also requires sophisticated skills not yet acquired 500,000 years ago. However, catfish were seasonally abundant to early East African hominids at least two million years ago (see Chapter 13).

Shellfish are the best source of these brain selective minerals because of the high concentration of these minerals, and the greater year round abundance, accessibility, and digestibility of shellfish compared to other food groups. The rich density of shellfish just below the water line in mud or sand and on rocks, and their full exposure at low tide makes shellfish accessible to all age groups. This is particularly exemplified by the shellfish gathering by both the Moken children of Indonesia and the Meriam children of Mer Island in the Torres Straits off Australia.

Eggs, fish and nuts are not quite as good sources of these minerals as shellfish. In addition, seasonal and vulnerability to climatic variability (drought) and competitors sometimes reduce their availability. Consequently, compared to shellfish, significantly higher amounts of nuts, fish and eggs are needed to meet the requirement for brain selective minerals.

Meat has moderately high concentrations and good bioavailability of zinc, copper and iron but is poor in selenium and iodine. Owing to the skills necessary to catch animals, as well as competition from other predators and scavengers, meat would have been a sporadic hominid food, at least initially. Fruit and vegetables are also seasonal so, despite their abundance, the markedly lower concentrations and poorer bioavailability of most minerals make fruit and vegetables less reliable sources of these nutrients.

Present day varieties of pulses can be good sources of these minerals. However, as with cereals, domesticated pulses contain *anti-nutrients* such as phytate and goiterogens, which significantly limit their suitability as sources of brain selective minerals. Like milk, pulses and cereals would have required a significant period of domestication to make them practical sources of these minerals.

Of course, no single food group would have been consumed to the exclusion of all others. Variety would have been as desirable and unavoidable in paleolithic times as it is now. A combination of shellfish, fish, nuts, eggs, meat and fruit seems probable. The key point is that *any* amount of shellfish would have helped considerably to meet the needs for these important minerals. Thus, occupation of shorelines would have provided unparalleled access to brain selective minerals while creating no dietary, physical or cultural disadvantage.

Interactions Between Brain Selective Minerals

Individually, iodine, iron, copper, zinc and selenium have important roles in human nutrition and in several aspects of brain structure and function. Specific deficiencies of these nutrients clearly have distinct pathology related to their precise roles in different processes or structures of the cell. These trace minerals also interact with each other, sometimes synergistically, sometimes in parallel, and sometimes competitively. This builds complexity into meeting the requirement for these nutrients and makes it more important for brain function that minimum levels of the whole cluster of brain selective nutrients be available.

For instance, iron and iodine are probably the most important nutrients for cellular respiration, eg. to fuel the 'expensive' organs like the brain and to prevent fatigue and weakness in continuously active, muscular organs like the heart. Iodine and iron act through different mechanisms but are complementary in this role. They are not alone in controlling the rate of cellular respiration in that each depends on another trace mineral for its activity. Iron cannot prevent anemia unless there is sufficient dietary copper, because copper is required for hemoglobin synthesis. Iodine cannot be into incorporated into triiodothyronine unless there is sufficient selenium for the deiodinase enzyme. Selenium is also involved in heme synthesis, thereby affecting the ability of iron to contribute to synthesis of hemoglobin.

Excess of some minerals inhibits absorption, metabolism or utilization of others. One of many such examples is that excess iron or copper inhibits absorption of zinc. Moderate to high intakes of some 'anti-nutrient' components of foods, i.e. phytate, are also serious nutritional problems because they inhibit absorption of most trace minerals. Plant-based goiterogens create a particularly serious and widespread problem for iodine absorption and utilization.

Brain Selective Minerals and Lipid Synthesis

In addition to their diverse roles in cellular biochemistry and energy metabolism, iodine, iron, zinc, copper, and selenium all have one broad metabolic effect in common that is directly relevant to brain function and evolution – they all interact importantly in lipid metabolism (Table 7.7). Lipids are major components of both the neuronal synapses needed to receive and integrate information leading to the development of a signal, and the myelin needed to protect the integrity of the signal being sent from the brain to the target tissue.

Table 7.7. Interactions of brain selective minerals in lipid metabolism.

1. Iron is a component of the desaturase enzymes required for the synthesis of long chain monounsaturated and polyunsaturated fatty acids.
2. Copper has a role in delta-9 desaturation, possibly as part of its function in the electron transport chain.
3. Iodine is needed for ketogenesis and normal myelination.
4. Zinc is required for delta-5 and delta-6 desaturase activity, possibly through its effects on electron transport. This effect makes zinc particularly important to ensure sufficient capacity to synthesize arachidonic acid and docosahexaenoic acid when these two fatty acids are not present in the diet.

Iron is a structural component of the *desaturase* enzymes, which are required to add double bonds to long chain fatty acids. There are several fatty acid desaturases that are structurally similar but not interchangeable because they insert double bonds in different locations in fatty acids. They all have in common a structural requirement for one iron atom per enzyme molecule. One of the desaturases, the *delta-9 desaturase*, makes the eighteen carbon monounsaturated fatty acid, *oleic acid*, from the eighteen carbon saturated fatty acid, *stearic acid*. The *delta-5 and delta-6 desaturases* are needed at separate steps in the synthesis of polyunsaturated fatty acids, i.e. *docosahexaenoic acid* from *alpha-linolenic acid* and *arachidonic* acid from *linoleic acid* (see Chapter 8).

Although there is intense discussion amongst nutritionists and pediatricians concerning how to get adequate docosahexaenoic acid into the developing human brain, there is actually a lot more oleic acid than docosahexaenoic acid in the brain. Oleic acid is used in all brain lipids, i.e. both in myelin and in synapses, whereas docosahexaenoic acid is principally found in synapses.

The crucial point about iron-dependent desaturation of fatty acids is that although there is abundant stearic and oleic acid in the diet and in body fat stores of all mammals, the brain is largely incapable of acquiring sufficient amounts of these two fatty acids from outside itself. The brain takes up small amounts of stearic and oleic acid but the amount is too low to meet the requirements of brain growth and lipid deposition. Thus, in contrast to the brain, other organs such as the liver can get their oleic acid directly from the diet or via synthesis from stearic acid.

Iron deficiency inhibits delta-9 desaturation in all tissues studied and hence affects tissue stearic and oleic acid levels, but organs other than the brain do not depend on synthesis to make oleic acid. However, iron deficiency reduces the oleic acid content of brain synapses and myelin. Iron deficiency also decreases the brain content of other longer chain monounsaturated fatty acids derived from oleic acid that, like oleic acid, are important components of myelin and contribute to optimal learning and memory.

Copper is also needed for delta-9 desaturation and oleic acid synthesis. Thus, copper deficiency also reduces oleic acid levels in tissues whereas copper excess increases oleic acid. Copper's exact role in desaturation is much less clear than for iron. It is not a structural component of the desaturase protein itself but copper does have a role in electron transport at the cytochrome C oxidase transferring electrons to the terminal step at the desaturase protein itself.

The main consequence of low brain oleic acid caused by iron or copper deficiency is hypomyelination causing weakness or errors in the electrical signal. The outcome of low desaturation caused by copper deficiency is therefore spasticity, tremor and muscle weakness and fatigue.

Iodine is also needed for normal myelination and at several stages in ketogenesis. Ketogenesis is essential not only to provide alternate fuel to the brain (ketones) but for the synthesis of lipids such as cholesterol and long chain fatty acids destined for both neuronal synapses and myelin. Hence, the role of iodine in ketogenesis impacts on myelin synthesis because long chain saturated (palmitic and stearic acids) and monounsaturated fatty acids (oleic acid) used in myelin synthesis are important products of ketogenesis in the developing brain.

The process of electron transport from the donor proteins (cytochromes) to the fatty acid desaturases is dependent on zinc. The delta-5 and delta-6 desaturases involved in converting linoleic acid and alpha-linolenic acid to the longer chain polyunsaturates seem somewhat more sensitive to zinc deficiency than does the delta-9 desaturase converting stearic acid to oleic acid.

As a result, synthesis of the long chain polyunsaturates, especially arachidonic acid, is markedly impaired by even mild to moderate zinc deficiency. This inhibition of the desaturases by zinc deficiency is so strong in infant humans and in young animals that it causes a more rapid decline in tissue arachidonic acid and docosahexaenoic acid than does the direct dietary deficiency of all the omega 6 or omega 3 polyunsaturated fatty acids.

***

Several key nutrients participate in the control of human growth, metabolism and brain function. Iodine promotes early development but it depends on zinc and amino acids for protein and tissue building. Iodine stimulates cellular metabolism but it depends on copper and iron for the cellular furnaces (mitochondria) to generate heat and maintain body temperature.

Some animals can survive in environments that are quite low in these required nutrients but they are all small-brained. For several required nutrients, humans now occupy a large number of environments that no longer meet their daily requirements for normal development. In industrialized countries, this problem has been artificially addressed by iodine supplements but nutrient deficiencies, principally for iodine, iron and zinc, remain a problem in many areas.

The metabolic interactions within brain selective minerals and between brain selective minerals and other nutrients represent some of the many such interactions between the full spectrum of essential nutrients needed for normal human development. This complexity and interdependence on nutrients coevolved with complex, multicellular systems and is widespread in both plants and animals; it is not a function of hominid or human evolution in particular.

Two (or even five) million years ago, some groups of hominids may have been genetically predisposed to having a brain weighing 1500 grams but it didn't happen. I believe that a hominid brain weighing 1500 grams didn't arise until 50-100,000 years ago because, prior to that, there was insufficient nutritional support for it. Even if only one nutrient is 'limiting' and all the others are available and working in concert, permanently changing key metabolic processes or limitations on cell structure, i.e. expansion of the cerebral cortex, cannot bypass dependence on that limiting nutrient.

Humans are now living beyond the optimal nutrient limits for intake of several nutrients. Adaptation will be necessary, either by making supplements more widely available or by moving back to the shorelines, or we will conceivably face evolutionary processes that could eventually reduce cognitive capacity.

Chapter 8

Docosahexaenoic Acid:
The Brain Selective Fatty Acid

Saturates, Monounsaturates, and Polyunsaturates

Fatty acids with no double bonds are *saturated*, meaning all the carbons in the chain except the two end ones have four neighbouring atoms each consisting of two carbons in the chain and two hydrogens sticking out to the sides. Each bond joins the carbons together at angles to one another but once ten or more carbons are lined up in a saturated fatty acid, the result is a carbon chain that is saw-toothed but, overall, is straight.

There are two alternatives to a fatty acid being saturated: The first alternative is that the fatty acid is *monounsaturated*, meaning that one pair of carbons, usually near the middle of the molecule, is linked together by a double bond. This effect of the double bond gives monounsaturates a kink in the middle, compared to saturates which are straight. Biologically, this makes a big difference to the way that fatty acids pack together; saturates jam together like sardines in a tin; monounsaturates figuratively have their elbows sticking out so fewer can be packed together.

The second alternative is that the fatty acid has more than one double bond. Two or more double bonds make the fatty acid *polyunsaturated*, which, in practice, means two, three, four, five or six double bonds. Using the elbows analogy, polyunsaturates have large elbows and become curved. At five or six double bonds, they even spiral into a helix.

Saturates pack together densely and so can become hard at room temperature (like butter or lard). Monounsaturates can pack a little so

151

they are unlikely to be hard at room temperature but the may become harder when is it colder, like in the refrigerator. Polyunsaturates are unlikely to harden except when very cold. Therefore, double bonds are a feature of major importance in the biology and chemistry of fatty acids.

Fatty acids common in the diet are mostly sixteen or eighteen carbons long. Those that are important in the structure of membranes are 16 to 24 carbons long. Fatty acids shorter than sixteen carbons are certainly present in the body but are uncommon in the diet and rarely important in membranes. The sixteen carbon fatty acids can be saturated (*palmitic acid*) or monounsaturated (*palmitoleic acid*). The eighteen carbon fatty acids can be saturated (*stearic acid*), monounsaturated (*oleic acid*), polyunsaturated with two double bonds (*linoleic acid*), or polyunsaturated with three double bonds (*alpha-linolenic or gamma-linolenic acid*).

Two fatty acids with the same chain length and same number of double bonds but with the double bonds in different places are called *isomers* because they are chemically similar but biologically quite different. Hence the two polyunsaturates with three double bonds each are both called linolenic acid but one is *alpha-linolenic acid* and the other is *gamma-linolenic acid*. This is the problem with fatty acids – their variety is enormous and multiplies not only as the chain length and the number of double bonds increase but also with the location of the double bonds in the molecule.

There are two key points about polyunsaturated fatty acid isomers: First, two isomers like alpha-linolenic acid and gamma-linolenic acid exist in the body but have very different functions. Second, one would expect that as the number of double bonds increases, so too would the number of isomers. Once there are three or four double bonds and 18 to 20 carbons, there are a large number of possible isomers but, biologically, the number of isomers is actually fairly limited. For example, there is effectively only one isomer of a polyunsaturate with four double bonds, and it has twenty carbons (*arachidonic acid*). For all intents and purposes, there is also only one twenty carbon polyunsaturate with five double bonds (*eicosapentaenoic acid*).

Of the many possible options, there are actually only two important isomers of polyunsaturates with 22 carbons and five double bonds. Most importantly, there is only one known isomer of the 22 carbon polyunsaturate with six double bonds – *docosahexaenoic acid* (Figure 8.1). Of all the possible carbon chain lengths and double bond isomers, arachidonic and docosahexaenoic acids are the two polyunsaturates most relevant to understanding brain function.

Metabolism of Omega 6 and Omega 3 Polyunsaturates

Polyunsaturates occur in two different families based on two different double bond positions (Figure 8.2). The naming is based on the location of the first double bond nearest the *omega* (or *methyl*) carbon, which is the end one with three hydrogen neighbours. The opposite end of a fatty acid is the *carboxyl* end because that is where the carbon-oxygen group (*carboxyl*) is located. One family of polyunsaturates has all its members with the first double bond starting at the sixth carbon from the omega carbon; hence, it is the family of *omega 6 polyunsaturates*. The other family is the *omega 3 polyunsaturates*.

Linoleic acid and alpha-linolenic acid are the 'parent' polyunsaturated fatty acids at the head of omega 6 and omega 3 families, respectively (Figure 8.1). In animals, a one step conversion turns the parent fatty acid into the next fatty acid in the series. Effectively, linoleic acid is converted through two intermediates to arachidonic acid, the primary 'long chain' omega 6 polyunsaturate in membranes.

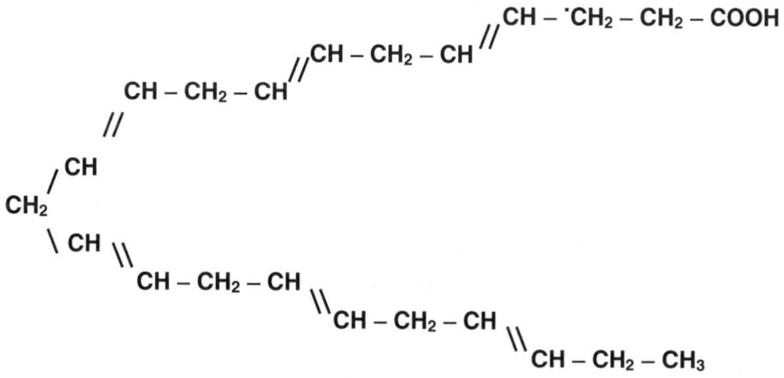

Figure 8.1. The structure of docosahexaenoic acid. Its unique combination of features includes having 22 carbons and six double bonds that are arranged at the 'omega 3' or methyl (CH_3) end of the molecule.

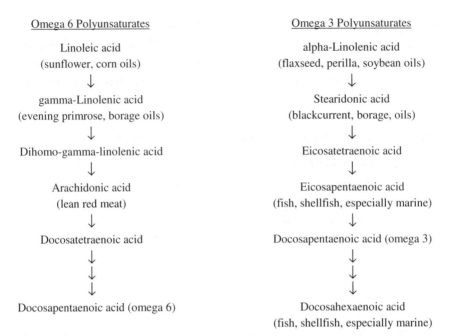

Figure 8.2. Metabolism of polyunsaturated fatty acids. Common dietary sources of these fatty acids are shown in brackets. Note that synthesis of the final fatty acids in both series require three steps that are not shown (hence, three consecutive arrows). Less than 5% of the overall metabolism of the parent polyunsaturates (linolenic and alpha-linolenic acids) proceeds through this pathway; the great majority is consumed via beta-oxidation (use as a fuel; see Figure 8.3). Not shown here are the four distinct families of *eicosanoids* (*prostaglandins*, *leukotrienes* and related substances) that are produced from dihomo-gamma-linolenic acid, arachidonic acid, eicosapentaenoic acid and docosahexaenoic acid.

For the omega 3 polyunsaturates, there is a similar sequence in which alpha-linolenic acid is converted through four intermediates on its way to becoming docosahexaenoic acid. With both the omega 6 and omega 3 polyunsaturates, it is a later fatty acid than the parent in the sequence that seems the most important for normal early development in babies, i.e. arachidonic acid in the omega 6 series and docosahexaenoic acid in the omega 3 series.

The fatty acid conversion pathway leading from linoleic acid to arachidonic acid and from alpha-linolenic acid to docosahexaenoic acid exists in humans and in animals. However, in most animal species the synthesis of long chain polyunsaturates is a slow process that is relatively tightly controlled. Thus, non-vegetarians can make a little arachidonic acid and docosahexaenoic acid but can also get some from the diet (meat and fish, respectively).

Polyunsaturates in Neonatal Body Fat Stores

Whether the mother has eaten sufficient of these two fatty acids or has made them herself, she passes more of them on to her fetus to help build its cell membranes and to store in its fat before birth than she stores in her own fat. That is why baby fat at birth contains more arachidonic acid and docosahexaenoic acid than does the mother's fat (Table 8.1). The logic of this seems to be that if the mother provides the baby with a ready-made supply of arachidonic acid and docosahexaenoic acid for its needs after being born, it will have a higher chance of developing normally because it will depend less on its immature capacity to make these long chain polyunsaturates. In effect the baby is better insured.

Table 8.1. Comparison of the fatty acid composition of body fat in human adults to that of body fat and brain in newborns. Adults accumulate much more linoleic acid and alpha-linolenic acid in body fat than do newborns or the newborn brain.

	Adult Body Fat	Newborn Body Fat	Newborn Brain
SATURATES:			
Palmitic Acid	20.1	43.0	28.5
Stearic Acid	4.7	4.1	18.2
MONOUNSATURATES:			
Oleic Acid	41.7	27.6	20.0
Omega 6 POLYUNSATURATES:			
Linoleic Acid	15.4	2.0	0.9
Arachidonic Acid	0.3	0.7	11.2
Omega 3 POLYUNSATURATES:			
α-Linolenic Acid	1.5	<0.1	<0.1
Docosahexaenoic Acid	0.1	0.4	8.4

Data compiled from Hirsch et al (1960), Sarda et al (1987), and Farquharson et al (1993).

The adult human body puts only small amounts of arachidonic acid and docosahexaenoic acid into body fat stores because these two fatty acids have important roles in cell membranes and there is very little of them left over to put in storage. Nevertheless, there is noticeably more docosahexaenoic acid and arachidonic acid in body fat at birth than in adults, which is an important difference between baby and adult human fat (Table 8.1).

There is a premium on docosahexaenoic acid and arachidonic acid in two ways; first, because they are needed for normal membrane structure and function and, second, because they are not made in abundant quantities from their parent fatty acids. Hence, storing 3 to 4 times more of these fatty acids in body fat at birth than in adults signifies the importance placed on these two fatty acids for normal infant development.

Much of human brain growth occurs after birth so being born with stores of key fatty acid nutrients makes good sense. If the mother stores too much arachidonic acid and docosahexaenoic acid in her much larger body fat stores, either the baby will have less insurance or the mother will then have to transfer them into her milk to make sure the baby gets the docosahexaenoic and arachidonic acid stores it needs. Hence, passing on as much of these two fatty acids as possible to the baby before birth also makes sense.

We recently showed that the typical healthy human baby is born with about three months store of docosahexaenoic acid and arachidonic acid in its fat. This is in addition to its own ability to make a little of these two fatty acids and to the normal daily delivery of these same fatty acids in the mother's milk. Hence, the healthy term baby's requirement for these two important long chain polyunsaturated fatty acids is normally quite well insured.

Fat contains three months supply of docosahexaenoic acid and arachidonic acid not because there is a high concentration of these fatty acids in baby fat but because there is normally about 500 grams of body fat at birth. The concentration of fatty acids in body fat of pre-term infants has not been reported but there would still be much less docosahexaenoic acid or arachidonic acid in total because pre-term infants have much less body fat than do term infants. Thus, the newborn premature human infant has less insurance to meet its energy needs and its needs for building up the lipid structure of all cell membranes, especially in the brain.

Though other species use arachidonic acid and docosahexaenoic acid in the brain and receive them in mother's milk, they do not store them in body fat at birth nor, indeed, even have visible body fat stores at birth. I would argue that they don't need to store these two fatty acids because their brains are not expanding very much after birth. The advantage of nutrient storage in body fat at birth is, in fact, a necessity as seen from the vulnerability to neurodevelopmental problems encountered by babies born prematurely, all of whom have low fat stores.

Docosahexaenoic Acid – The Brain Selective Fatty Acid

More has been written about the possible role of docosahexaenoic acid in human brain evolution than for any other brain selective nutrient. *'Fish is brain food'* is an adage that substantially predates the discovery in fish of the omega 3 polyunsaturated fatty acids that are now known to be important in brain function. Docosahexaenoic acid is the main polyunsaturated fatty acid in fish from marine temperate or polar waters; it is always one of the main polyunsaturated fatty acids in fish, whether from temperate fresh or salt water (Table 8.2), East African lake fish (Table 8.3) or in shellfish (Table 8.4).

Despite its similarity in structure to docosahexaenoic acid, omega 6 docosapentaenoic acid does not overcome the symptoms of docosahexaenoic acid deficiency. The same is even true of the immediate precursor to docosahexaenoic acid, *omega 3 docosapentaenoic acid.* Thus, the two fatty acids most similar to docosahexaenoic acid but containing only five double bonds (omega 6 and omega 3 docosapentaenoic acids) cannot effectively substitute for docosahexaenoic acid's six double bonds, thereby making docosahexaenoic acid alone a *brain selective fatty acid.*

Before turning to a description of the main factors affecting docosahexaenoic acid synthesis and why they would make a dietary source of docosahexaenoic acid essential for human brain evolution, it is notable that there is wide agreement amongst specialists that specifically docosahexaenoic acid has an irreplaceable role in brain lipid and photoreceptor structure. Thus, brain function and vision are compromised when docosahexaenoic acid is reduced and docosahexaenoic acid cannot be replaced in these roles by any other (even similar) polyunsaturated fatty acids. Despite this wide agreement today, this was not the case even ten years ago so the consensus has been slow to emerge.

Table 8.2. Docosahexaenoic acid in fish (milligrams/100 grams) from temperate fresh or salt water.

Fresh water:	Carp		114
	Catfish		234
	Northern Pike		74
	Walleye		225
Salt water:	Bluefish		519
	Cod		130
	Grouper		220
	Haddock		126
	Halibut		292
	Herring,	Atlantic	862
		Pacific	689
	Mackerel,	Atlantic	1400
		King	177
		Spanish	1010
	Mullet		108
	Perch, Ocean		211
	Pollock, Atlantic		350
	Smelt		536
	Tuna,	Bluefin	890
		Skipjack	185
	Turbot		880
Mixed fresh and salt water:	Salmon,	Atlantic	1115
		Chinook	944
		Chum	394
		Coho	656
		Pink	586
		Sockeye	653

Sophisticated techniques that assess how the exact shapes of molecules give them biological activity suggest that docosahexaenoic acid's shape is uniquely important to its function in the brain. The double bonds in docosahexaenoic acid each twist the molecule a little bit. With six double bonds, each twist helps contribute to a total curvature of about one turn of a helix or coil. This gives docosahexaenoic acid a somewhat cylindrical shape that probably varies in length, i.e. it can compress or lengthen a little bit like coiled spring with about one full turn. Other fatty acids with fewer double bonds cannot make the same coiled shape, which seems to be the factor that makes them unable to replace docosahexaenoic acid.

Table 8.3. Docosahexaenoic acid (milligrams/100 grams) in East African lake fish (modified from Broadhurst et al, 2002).

Mbelele (catfish; Lake Nyasa)	842
Carp (Njenu; Lake Nyasa)	363
Mfui (Lake Nyasa)	200
Kambale (Lake Nyasa)	227
Tilapia (Lake Turkana)	343
Perch (Lake Turkana)	447

Table 8.4. Docosahexaenoic acid (milligrams/100 grams) in shellfish.

American clam	420
Quahog clam	370
Surf clam	370
Crayfish	38
Blue crab	150
Queen crab	113
Blue mussels	253
Eastern oysters	292
Pacific oysters	250
Shrimp	222

Progress on understanding why docosahexaenoic acid's shape is biologically important is advancing most rapidly in studies on its role in the photoreceptor of the mammalian eye where docosahexaenoic acid interacts with *rhodopsin* in transforming photons of light into electrochemical signals sent to the brain. In a manner analogous to its interaction with rhodopsin, docosahexaenoic acid is also believed to be important in the transfer of signals across the synapse between two nerve cells. How this happens is still unclear. What is clear is that those human infants or animals with low levels of brain docosahexaenoic acid have poorer memory and slower learning.

Capacity to Make Docosahexaenoic Acid

Docosahexaenoic acid is the eighth in the sequence of omega 3 polyunsaturated fatty acids (Figure 8.1). If they are provided with the precursor omega 3 fatty acid, alpha-linolenic acid, all mammals in which this pathway has been studied can make some docosahexaenoic acid. The efficiency of the desaturation – chain elongation pathway varies markedly across species, with most rodents being more efficient than humans but carnivores such as cats and lions being less efficient than humans. Caution is therefore needed in extrapolating from one species to another, especially from rats, which are a common laboratory model for human physiology and biochemistry (this is less of a problem with lions).

Humans have a low capacity to make docosahexaenoic acid but that capacity does exist. Of the seven omega 3 polyunsaturated fatty acids ahead of docosahexaenoic acid, two are found in the diet. For unknown reasons, the other five omega 3 polyunsaturated fatty acids in the sequence before docosahexaenoic acid occur in only minute amounts in animals and are not found at all in plants.

The two that are found in the diet are alpha-linolenic acid and eicosapentaenoic acid (twenty carbons long and containing five double bonds). Apart from a few rare fungi and lichens, alpha-linolenic acid is the only omega 3 polyunsaturated fatty acid in terrestrial plants. It appears to be present in all plants, especially in the leaves, and can account for as much as 60% of the plant's fatty acids. However, with the exception of the seeds, most plant tissues contain very low levels of fat so large amounts of plant material would have to be consumed for it to be considered a good source of alpha-linolenic acid.

Alpha-linolenic acid is also present in animal fat in proportion to its intake so it is generally higher in herbivores consuming it directly than in carnivores, which only get it when they eat herbivores. Like docosahexaenoic acid itself, eicosapentaenoic acid is common only in fish, with more being found in cold water marine fish than in tropical or freshwater fish.

The reason that pre-formed dietary docosahexaenoic acid was probably needed for human brain evolution is that the simple presence of the biochemical pathway to convert alpha-linolenic acid to docosahexaenoic acid does not guarantee sufficient capacity to synthesize docosahexaenoic acid to meet the brain's needs. This is most clearly evident in infants given a formula containing no docosahexaenoic acid. Until two years ago (ten years ago in Europe), neither the USA nor Canada allowed infant formula manufacturers to put docosahexaenoic acid in their infant nutrition products. This was despite that fact that regardless of high or low docosahexaenoic acid intake, it is present in the breast milk of mothers around the world.

The main reason for banning docosahexaenoic acid from milk formulas was a fear that if it were added, its six double bonds would cause it to go rancid (like in decaying fish). Additionally, several studies had shown that infants could make docosahexaenoic acid, so it was assumed that the risk of rancidity combined with the ability to synthesize docosahexaenoic acid made it redundant in milk formulas.

At the time, little thought was given to whether the infant had sufficient capacity to make docosahexaenoic acid. A couple of papers published during the past decade each showed that infants on a formula not containing docosahexaenoic acid do not achieve the same brain level of docosahexaenoic acid as breast fed infants, i.e. they cannot make enough docosahexaenoic acid to achieve 'normal' brain docosahexaenoic acid levels. These papers were instrumental in contributing to reversing the ban on docosahexaenoic acid in milk formulas. They were a powerful reminder that the presence of a biochemical pathway does not insure that, starting with the main precursor (alpha-linolenic acid), the end product of that pathway (docosahexaenoic acid) can necessarily be made in sufficient amounts.

Like other enzyme pathways, the desaturation and chain elongation of alpha-linolenic acid (or linoleic acid) depends on cofactor nutrients including vitamin B_6, magnesium, iron and zinc. This means that if one or more of these cofactors is not present in sufficient amounts in the diet, docosahexaenoic acid synthesis will be impaired. Indeed, if animals or humans are undernourished, their need to burn (oxidize) fatty acids increases and net docosahexaenoic acid synthesis ceases. Therefore, the desaturation – chain elongation pathway cannot guarantee that docosahexaenoic acid will be reliably synthesized unless essentially all nutritional circumstances affecting this process are consistently favorable.

In other words, in humans, there is enough of a bottleneck at one or more points in the pathway that little or no end product gets made. That is why, regardless of the intake of alpha-linolenic acid, the infant brain has lower docosahexaenoic acid if docosahexaenoic acid isn't provided in the milk.

Veganism and Docosahexaenoic Acid Insurance

Vegans eat strictly plant-based diets. Unless they take supplements containing docosahexaenoic acid, they consume next to no docosahexaenoic acid. Marine or shore-based plants may contain eicosapentaenoic acid but no edible plants provide docosahexaenoic acid. Vegans may still be able to synthesize sufficient docosahexaenoic acid for normal development but they have much less margin of error than in individuals with even a modest dietary intake of docosahexaenoic acid.

Even with abundant intake of alpha-linolenic acid, the desaturation – chain elongation pathway is an inefficient way to make the amount of docosahexaenoic acid necessary for the brain. With the exception of shore based plants like dulce and some seaweeds, alpha-linolenic acid is the only omega 3 polyunsaturated fatty acid in plants. As such, it follows that a strictly plant-based diet is an inefficient way to acquire docosahexaenoic acid.

Norman Salem's group at the National Institutes of Health in Rockville, Maryland has shown that, even with their more efficient synthesis of docosahexaenoic acid, suckling rats still do not achieve optimal brain docosahexaenoic acid levels unless docosahexaenoic acid is actually in the milk. This strongly implies that *veganism* a risky way to achieve normal brain development in animals as divergent as rats and humans. That is not to say it is impossible, just more difficult. The point is that if it were hard to sustain normal brain development when docosahexaenoic acid is limiting, ultimately, it would be even harder to expand brain size and cognitive capacity without an even more reliable dietary source of docosahexaenoic acid.

It is impossible to achieve normal brain function in the absence of both body stores of docosahexaenoic acid at birth and a lack of docosahexaenoic acid in the diet. On the one hand, vegans consuming no foods of animal origin and, hence, consuming no dietary docosahexaenoic acid (which is present in low amounts in milk, cheese or eggs) have not been reported to be cognitively impaired. They have lower docosahexaenoic acid in milk and in the infants' blood but their infants are believed to have normal development.

Thus, marginally lower docosahexaenoic acid *per se* is not necessarily catastrophically limiting for brain function. Many millions of human infants have been raised on formulas containing no docosahexaenoic acid. Perhaps not all of them were high achievers but some undoubtedly were, so it is possible to achieve normal brain function without a dietary source of docosahexaenoic acid.

Although docosahexaenoic acid is lower in vegans than in those consuming some fish or meat, it is not as low as when there is no dietary source of any omega 3 polyunsaturated fatty acid. Thus, vegans must be able to make some docosahexaenoic acid and may well have somewhat higher efficiency of docosahexaenoic acid synthesis because there is no dietary docosahexaenoic acid to inhibit its own synthesis by negative feedback. Their infants don't achieve the tissue docosahexaenoic acid levels of infants consuming some docosahexaenoic acid but they may still be above the necessary threshold for docosahexaenoic acid adequacy for the brain.

Clearly, it is more difficult to achieve optimal tissue docosahexaenoic acid without docosahexaenoic acid in the diet; dietary docosahexaenoic acid is invaluable insurance for acquisition of optimal brain docosahexaenoic acid and optimal brain function. As with the deficiency symptoms for other nutrients, there is no specific threshold for docosahexaenoic acid deficiency symptoms involving vision or brain function. Rather, the effects vary with age, species, the test used, etc. Thus, we cannot say that unless *H. erectus* got a precise amount of docosahexaenoic acid, *H. sapiens* would not have evolved.

The point in relating dietary docosahexaenoic acid to human brain expansion is not that an exact amount of docosahexaenoic acid or any other single nutrient was needed to have expanded the hominin brain. Rather, at some point, lower intake or synthesis of docosahexaenoic acid limits brain function sufficiently that expansion and increased complexity of the brain is much less likely when docosahexaenoic acid intake is low. In that sense, there is a threshold docosahexaenoic acid intake for optimal human brain function.

Zellweger Syndrome

Beyond the issue of nutritional deficiencies that inhibit docosahexaenoic acid synthesis is the rarer but much more serious problem of genetic mutations blocking docosahexaenoic acid synthesis. The last stage of docosahexaenoic acid synthesis requires organelles in the cell called *peroxisomes*. In addition to other serious consequences, a disorder in which peroxisomes are absent almost totally inhibits docosahexaenoic acid synthesis. *Zellweger syndrome* is one of a cluster of related *peroxisomal biogenesis disorders* that are characterized by extreme mental retardation and early death.

In general, this disease is less severe before weaning from breast milk because the mother is still providing pre-formed docosahexaenoic acid in her milk. Once on a milk formula, most of which still do not contain docosahexaenoic acid, the absence of incoming docosahexaenoic acid as well as the inability to make it essentially prevents further neurological or physical development in Zellweger cases.

Manuela Martinez from the University of Barcelona has spent many years characterizing the involvement of docosahexaenoic acid in Zellweger syndrome. She has clearly demonstrated both the catastrophically reduced brain levels of docosahexaenoic acid in these infants and a modest beneficial effect of supplemental dietary docosahexaenoic acid on development in Zellweger cases. Though Zellweger syndrome is relatively rare, the point of describing it (or, indeed, the nutritional cofactors needed for efficient docosahexaenoic acid synthesis) is that depending on docosahexaenoic acid synthesis is a much less reliable way to achieve necessary tissue docosahexaenoic acid levels than consuming pre-formed docosahexaenoic acid in the diet.

Arachidonic Acid – Important but not Brain Selective

Arachidonic acid is an omega 6 polyunsaturated fatty acid that in some ways is structurally and metabolically similar to docosahexaenoic acid. However, the two are sufficiently distinct that arachidonic acid cannot be classified as a brain selective fatty acid. The human brain content of arachidonic acid is broadly equivalent to that of docosahexaenoic acid; in some brain lipid classes it is higher, while in others it is lower. With twenty carbons and four double bonds, arachidonic acid is neither as long nor as unsaturated as docosahexaenoic acid.

In general, the biological function of arachidonic acid has been studied for longer and its metabolism is arguably better understood that that of docosahexaenoic acid. Nevertheless, compared to docosahexaenoic acid, less is known about what arachidonic acid actually does in the brain.

Arachidonic acid is the main precursor to a myriad cascade of bioactive molecules collectively called *eicosanoids*. Eicosanoids include the *prostaglandins*, *leukotrienes* and several related substances. Eicosanoids have diverse and sometimes confusing effects on immune function, vascular tone, platelet aggregation and many other processes. They are regarded as key signalling molecules between cells but, in excess, i.e. in inflammation, can turn on the body and do considerable damage. Inflammatory joint disorders like arthritis are treated by aspirin and other drugs which block the pain and swelling caused by excess eicosanoid synthesis.

Eicosanoids can be produced in the brain from arachidonic acid but whether they are essential to brain biochemistry is unknown. Whether the relatively high concentration of arachidonic acid in the brain has any direct bearing on brain eicosanoid production is also unknown.

In contrast to docosahexaenoic acid, it is relatively hard to increase or decrease brain arachidonic acid levels. Raising and lowering tissue levels of a molecule normally present in the body is a useful tool in determining what that molecule does. Arachidonic acid is primarily found in meat and fish so if these foods are not present in the diet, as with docosahexaenoic acid, some arachidonic acid can be made from its main precursor, linoleic acid, which is widely distributed in foods. Nutritional limitations apply to the synthesis of arachidonic acid from linoleic acid in the same way they apply to docosahexaenoic acid synthesis.

Providing linoleic acid as the only dietary omega 6 polyunsaturated fatty acid does not reduce brain arachidonic acid, even though, in the analogous experiment, brain docosahexaenoic acid is decreased when diets are provided in which alpha-linolenic would be the only dietary omega 3 polyunsaturated fatty acid. The same desaturation – chain elongation enzymes are used to make both arachidonic acid and docosahexaenoic acid but fewer steps are needed to make arachidonic acid from linoleic acid than to make docosahexaenoic acid from alpha-linolenic acid. Arachidonic acid synthesis also does not need to pass through peroxisomes.

In experimental animals and in human infants, even a severely deficient intake of linoleic acid reduces brain arachidonic acid much less than the reduction in brain docosahexaenoic acid that occurs with deficient intake of alpha-linolenic acid. Compared to brain docosahexaenoic acid, either it is somewhat easier to make brain arachidonic acid or it is somewhat easier to prevent it from being broken down and removed from the brain. Clearly, despite both being polyunsaturates that are required in the brain, arachidonic acid and docosahexaenoic acid are not metabolized by the body in exactly the same way.

Brain levels of arachidonic acid are much less vulnerable to a dietary deficiency of omega 6 polyunsaturates than are brain levels of docosahexaenoic acid when intake of omega 3 polyunsaturates is low. The point is that, compared to docosahexaenoic acid, arachidonic acid seems more tenaciously held by the brain. This had made brain arachidonic acid more difficult to manipulate and study but, equally, is an indication of its importance in the brain.

Since it is almost impossible to deplete brain arachidonic acid even using diets not containing arachidonic acid, a dietary source of arachidonic acid seems less important for brain development than a dietary source of docosahexaenoic acid. Nevertheless, arachidonic acid is present in shore-based foods and in meat so, in contrast to docosahexaenoic acid, it is possible to meet arachidonic acid requirements either from the diet or via synthesis. Meat alone could supply the body's arachidonic acid requirements but is unlikely to be able to meet the body's docosahexaenoic acid requirements. Shore-based foods meet the requirements for both these fatty acids. Hence, dependence on terrestrial foods (meat or plant material) would be sufficient for arachidonic acid but would be insufficient for docosahexaenoic acid.

Other Important Fatty Acids in the Brain

Arachidonic acid and docosahexaenoic acid are unquestionably important in brain development. Nevertheless, in some ways, they are actually no more important than other lipids present in brain. Brain phospholipids need relatively large amounts of at least one saturated fatty acid, *palmitic acid*, and even larger amounts of the principal monounsaturated fatty acid, *oleic acid*.

One new and exciting role for palmitic acid recently discovered by Philip Beachy's group at Johns Hopkins University in Baltimore is as an obligatory component for the normal expression of at least one of the family of proteins called *hedgehog*. Hedgehog proteins help guide the normal development of the embryo, especially its brain. Under- or over-expression of hedgehog is catastrophic for normal development.

Cholesterol is one key lipid constituent needed for hedgehog expression. Beachy's recent discoveries show that supplying just the right amount of palmitic acid and cholesterol is important for this protein's expression. Too much or too little palmitic acid or cholesterol and hedgehog expression is altered and development becomes distorted.

This is one example of why it is as important to the brain to get the supply of cholesterol and palmitic acid right as it is to get the supply of arachidonic acid and docosahexaenoic acid right. The brain has evolved the ability to regulate its lipid composition very carefully and to do so has retained the ability to make most fatty acids and cholesterol. That way it is independent of the vagaries of the dietary supply of palmitate (or cholesterol). The brain still depends on the diet for the long chain polyunsaturates, somewhat analogous to vitamins or indispensable amino acids.

Oxidation of Linoleic and alpha-Linolenic Acids

Since both the mother and the developing fetus or infant can make docosahexaenoic acid and arachidonic acid, why is it important to have them stored as insurance in body fat? The reason is that there is just one chance, one critical period, to get brain development correct. Each in the series of critical windows in brain development depends upon successful completion of the previous one; otherwise, John Dobbing's 'deficits and distortions' start to arise and learning ability becomes compromised (Chapter 4). The bigger the brain and the more it develops postnatally, the higher the fuel and structural needs and the greater the vulnerability if they cannot be met. Therefore, having two of the fatty acids needed by the brain pre-formed and ready in fat stores reduces the risk that problems in their synthesis will compromise early brain development.

If the parent polyunsaturates (linoleic acid and alpha-linolenic acid) can be converted to arachidonic acid and docosahexaenoic acid, respectively, why is it important to have stores of arachidonic acid and docosahexaenoic acid? Their stores are important because the synthesis pathway is susceptible to inhibition – it isn't dependable. Indeed, it isn't even known what the capacity of the pathway actually is except that measurable amounts of conversion occur. The pathway can be stimulated by some hormones, inhibited by others and is particularly vulnerable to inhibition by low intake of certain cofactor nutrients.

More importantly, the amount of conversion depends on availability of linoleic and alpha-linolenic acids. The problem is that linoleic and alpha-linolenic acids are both extensively oxidized in the body as fatty acid fuels (Figure 8.3). It may seem illogical to take a vitamin-like fatty acid such as alpha-linolenic acid and use it as a fuel when there are plenty of other fatty acids that could be fuels, yet none of these others can be converted to docosahexaenoic acid. How could such a risk be taken during evolution?

Looked at the other way around – oxidation of alpha-linolenic acid would not be a big risk if the maternal diet could be relied upon to provide a constant and abundant pre-formed source of docosahexaenoic acid, i.e. from shellfish, fish or other shore-based foods. In fact, relying on a multi-step enzyme pathway that responds negatively or positively to various hormones and can be inhibited by dietary deficiency of iron, zinc, magnesium or vitamin B_6 is actually a lot more risky than relying on an abundant dietary source of pre-formed docosahexaenoic acid. When the mother consumes docosahexaenoic acid, she stores some in her fat, transfers some to the fetus, and transfers some to the suckling baby via her milk.

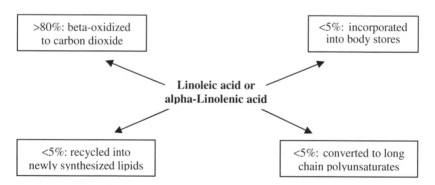

Figure 8.3. Overall utilization of the parent polyunsaturates, linolenic and alpha-linolenic acid. Unlike Figure 8.1, this shows the extensive catabolism of these two fatty acids, resulting in their use as fuels or in the synthesis of other lipids. The percents shown are compiled from various studies and represent values that are generally seen in a wide variety of healthy, physiological states.

These various ways of getting pre-formed docosahexaenoic acid into the baby actually appear to make its synthesis in the baby almost redundant. In this situation, the presence of the pathway to make docosahexaenoic acid becomes an evolutionary vestige – needed only by animals with small brains, no body fat at birth and no dietary docosahexaenoic acid, but not by those consuming and storing docosahexaenoic acid.

If the baby doesn't really need to make docosahexaenoic acid, it can afford to oxidize alpha-linolenic acid. Thus, as precursors to make long chain polyunsaturates, linoleic acid and alpha-linolenic acid are actually *redundant* in those species consuming a shore-based diet. In diets that are not shore-based, linoleic acid and alpha-linolenic acid would not be redundant but then the organism's brain development has to be scaled back to stay within the constraints of low and variable synthesis of arachidonic acid and docosahexaenoic acid, respectively.

Low Levels of Linoleic and alpha-Linolenic Acids in the Brain

As part of our research into the magnitude of oxidation of linoleic acid and alpha-linolenic acid, we undertook studies in which these fatty acids were 'labelled' with a marker or tracer such as carbon-13 or carbon-14. The metabolism of these tracers was then assessed by quantitative methods involving nuclear magnetic resonance spectroscopy and mass spectrometry. To our great interest and surprise, a considerable amount of the oxidized carbon from linoleic acid and alpha-linolenic acid appeared in newly synthesized lipids like cholesterol. This occurred most commonly in infant animals but occurs under all circumstances we have studied.

In fact, other laboratories, notably that of Tom Brenna at Cornell University, have confirmed our observations. Indeed, Andrew Sinclair of the Royal Melbourne Institute of Technology in Australia demonstrated the same effect thirty years ago but his observations remained unnoticed until we picked up on them and extended his work early in the 1990s. Further work we have done since then has shown that, to end up in a newly synthesized lipid, the carbon from oxidized linoleic or alpha-linolenic acid is recovered in ketone bodies which are then incorporated into the newly made lipids.

Making new fatty acids out of linoleic or alpha-linolenic acid seems like a complicated and wasteful use of two valuable fatty acids. However, the fact is that the brain and body fat of infants contains very low amounts of linoleic acid and alpha-linolenic acid. Very low means linoleic acid is under 2% of total fatty acids and alpha- linolenic acid is almost undetectable (Table 8.1).

In contrast, in fat from adults, linoleic acid is typically 10-15% of the total fatty acids present and alpha-linolenic acid is 0.5-2% of the total. These amounts vary with the type and amount of fat one eats but are reasonable as a normal range. Hence, adults usually have five to ten times more linoleic acid and α-linolenic acid in their body fat than do infants. Interestingly, in adult or infant brain, linoleic acid and alpha-linolenic acid are also present in very low amounts - about 1% and 0.1% of total fatty acids, respectively.

It isn't a coincidence that the body fat and brain of infants have similarly low amounts of linoleic and alpha-linolenic acids. Nor is it a coincidence that infant and adult brains contain similarly low amounts of these two fatty acids. In my view, the infant's body fat is helping supply the fatty acid needs of the brain and so it's body fat contains polyunsaturates according to the needs of the brain during its early development. Compared to other fatty acids, there isn't much linoleic or alpha-linolenic acid in the brain and yet two polyunsaturates are needed in the brain, just not linoleic or alpha-linolenic acid.

Hence, linoleic acid and alpha-linolenic acid actually seem less important in the developing infant than they are in the adult. A greater dependence by the infant on ready-made docosahexaenoic acid and arachidonic acid frees up the linoleic acid and alpha-linolenic acid for other purposes. Ketone body synthesis is a beneficiary of the relaxed need by the infant for linoleic and alpha-linolenic acid to make the longer chain polyunsaturates.

Different Strategies to Obtain Fatty Acids

The mammalian brain developed three distinct strategies to obtain an appropriate mixture of fatty acids for its membranes (Table 8.5). The first strategy was to make the fatty acids it needed, i.e. palmitic acid, stearic acid, and oleic acid, and ignore the presence of these fatty acids in the diet. This strategy was more biochemically and metabolically expensive but it also gave the best guarantee for getting exactly the required amount of the fatty acids that were needed. If only this strategy had been employed in brain evolution, brains would have remained small and simple.

Table 8.5. Three different strategies evolved by the mammalian brain to acquire fatty acids.

1.	Synthesis within the brain: This strategy is used mainly for palmitic, stearic and oleic acids, which are essential in the brain, and makes brain entirely independent of external sources of these fatty acids.
2.	Combination of *efficient* synthesis and *highly regulated* transport: This strategy is reserved for arachidonic acid, thereby guaranteeing appropriate brain levels under most conditions.
3.	Combination of *inefficient* synthesis and *moderately efficient* transport. This strategy is reserved for docosahexaenoic acid, thereby usually guaranteeing appropriate tissue levels. Nevertheless, in the absence of dietary docosahexaenoic acid, human infants cannot acquire normal brain levels of docosahexaenoic acid, so this strategy leaves humans dependent on a dietary source of docosahexaenoic acid.

A second strategy was reserved for arachidonic acid. Like with other fatty acids, arachidonic acid is needed for brain membrane structure but arachidonic acid is different in having the additional responsibility for the delicate business of eicosanoid production. Therefore, more active capacity to transfer arachidonic acid across the blood-brain barrier was retained but the body also reserved the right to make sufficient amounts of arachidonic acid to meet the needs of the brain. Like with the fatty acids in the first strategy, this makes it difficult to deplete the brain of arachidonic acid. Unlike the other fatty acids, arachidonic acid fairly easily accesses the brain from the diet but the brain is not dependent on the diet for arachidonic acid. This two-pronged strategy (synthesis and transfer from the diet) was used only once and seems reserved for arachidonic acid.

The third strategy was to remain dependent on the diet for docosahexaenoic acid. The brain of no mammalian species has escaped the vitamin-like nature of docosahexaenoic acid. Those that couldn't obtain docosahexaenoic acid in the diet were prevented from brain expansion. Those that could reliably obtain docosahexaenoic acid were able to meet the requirement for the only dietary fatty acid that limited brain expansion. Note that there is no strategy to acquire linolenic or alpha-linolenic acid by the brain. Brain levels of these two fatty acids are very low and they appear unimportant for brain function.

There are only two mammalian species widely recognized to have both disproportionately large brains and advanced cognition – the bottle-nosed dolphin and humans: Both employ these three strategies to obtain brain fatty acids and both depend on docosahexaenoic acid as a brain selective fatty acid. By remaining in the marine environment, only the dolphin has retained liberal access to docosahexaenoic acid (and other brain selective nutrients).

Part 2

The Shore-Based Scenario

Chapter 9

Genes, Brain Function and Human Brain Evolution

Genes and Biological Complexity

In general terms, there are three basic options by which genes influenced human brain evolution. Each option involves at least one gene but more likely a gene cluster that works in sequence or in concert. First, a gene or gene cluster promoting brain development started to become functional in hominids but did not exist in a functional state in primates. Second, a gene or gene cluster suppressing brain development ceased to be functional in hominids but remained active in primates. Third, a combination of these two promoter and suppressor options occurred.

The promotion or suppression of brain development by a mutated gene can be direct or indirect. No one knows exactly which genes control brain development or how but, for argument's sake, let's take a hypothetical case in which a gene controlling brain cell production (*neurogenesis*) is changed in a way resulting in a longer period of neurogenesis. Longer neurogenesis results in more brain cells, which results in a bigger brain. By affecting, say, the replication process in neuroblasts that determines neuron number, the gene controlling neurogenesis can act directly on the final steps in that control process. A mutated gene could also affect production of a master hormone like thyroxine, which controls several aspects of fetal or childhood development by influencing expression of many genes.

The point is that at least two types of strategies are needed to examine the evolution of differences in complex organs like the brain: On the one hand it is essential to be focussed in order to understand how a specific gene works and build up a framework one gene and one species at a time. Whether that single gene works the same way in humans as in genetically related but cognitively different primates then needs to be determined. Humans have a large number of genes in common with species as different as houseflies and even bananas, so using other species to model what human genes do may be informative but is also limited in utility.

On the other hand, it is also essential to respect the fact that evolution involves whole organisms and not single genes, single proteins, or single cells. Humans are large organisms and their brains are perhaps the most complex structures in biology. Most of the body plan of humans and chimpanzees is very similar, with the differences being largely restricted to skeletal structure, body hair development, skin pigmentation, neonatal fatness and brain size. A 1% difference between the genomes of humans and chimpanzees still represents about 300 genes. Some of those gene differences may be silent, while others may be more affected by environmental differences in humans than in chimpanzees, or *vice versa*.

Genes controlling brain complexity and body fatness in neonates are likely to be functionally linked because humans lacking neonatal fat (premature infants), have a higher than normal risk that full development of brain complexity will not be achieved. In fact, the very existence of a risk that brain complexity will fail is an important clue to examining genetic and environmental controls on that risk of failure. Unlike the brain, the heart or lungs, or most other organs for that matter, have a low risk of failing to develop normally. Certainly, their development seems much less vulnerable to nutritional and other environmental insults than that of the brain. This is a clue that will need to be followed up in order to understand the close relationship between vulnerability and complexity in evolution of the human brain.

Amidst the hoopla surrounding the announcement of 'completion' of sequencing the human genome was a certain uneasiness that humans had fewer genes than had always been expected for our size and cognitive complexity. After all, the human body contains more than a hundred trillion cells, one hundred billion of which are in the brain. The laboratory roundworm has less than a thousand cells in its entire body, of which three hundred are in its nervous system. How could we have only about 30,000 genes whereas the lowly worm has almost 20,000 genes? Surely human complexity is dependent on more than a 50% increase in the number of genes compared to a worm.

The issue of how biological complexity resides in a relatively small increase in the number of genes is a thorny one forcing scientists to examine more closely how genes interact and are controlled. The number and complexity of proteins is a function both of the way genes are transcribed and of the interaction between genes, each of which can vary substantially and independently of the actual number of genes.

Genetic Proximity of Humans to Other Primates

The amino acid sequence of genes for brain proteins differs about twice as much between humans and other primates as do the genes for liver or white blood cell proteins. Svente Paabo's group at the Max Planck Institute has used this feature to explore the genetic proximity of humans to other primates. They studied 21,000 DNA sequences representing 17,000 genes in white blood cells, liver cells and brain cells. Each DNA coding sequence studied contained about a thousand base pairs so that both the differences and similarities of these genes were maximized.

For both white blood cells and liver, these DNA sequences were more similar between humans and chimpanzees than either was to macaques. However, in the brain genes, the chimpanzee sequences were more similar to those of the macaque than to the human. This greater difference between brain compared to liver or white blood cell genes was viewed as having arisen due to about a five fold acceleration in the rate of change of gene expression levels in the brain in hominids or humans themselves.

Paabo also noted that differences in the method for measuring how gene expression changes greatly affected the apparent rate of change of brain genes between humans and chimpanzees. Though still controversial, the apparently larger difference in gene expression in brain compared to other organs suggests that sometime in the 'recent' past allowed the human brain began to evolve faster than the chimpanzee brain.

Offsetting the enthusiasm with which genome differences are being pursued to establish evolutionary proximity of humans to other primates is the recognition that various mice strains of similar genetic diversity to humans and chimpanzees have differences in gene expression in the brain equivalent to those seen between humans and chimpanzees. In fact, even very similar inbred strains of mice, i.e. that differ in phenotype but which still most definitely represent one species, have equivalent variety in their genomes as do humans and chimpanzees. Hence, it is difficult to directly relate differences in DNA sequence to species differences.

It is now becoming clear that DNA sequence of the genome is only a crude framework for evolutionary distance between two species. Rather than differences in DNA sequence, the goal of more accurately determining when species split may become more accessible by studying differences in mRNA expression. mRNA codes for both qualitative differences between proteins, i.e. amino acid composition determining which protein is expressed, and for the amount of each protein expressed.

The types of proteins in the human and chimpanzee brain differ by 7-8%. A similar difference in brain proteins has been found in different mice strains. However, there was nearly a four-fold difference in total amount of some proteins in the human compared to chimpanzee brain, a difference that markedly exceeded what was seen across mouse strains. The question then arises as to whether differences in the various types or actual amount of proteins is more important in demarcating differences in brain size and function across species.

Despite problems with interpreting the results of molecular technology, it is becoming clear that the human brain differs in mRNA and protein expression compared to our nearest evolutionary relations. Despite seemingly small differences in the chimpanzee and human genome, where they differ most seems to be accompanied by significant differences in DNA or mRNA or both. How these differences translate into the marked cognitive differences between humans and chimpanzees still needs extensive work. It is also where environmental factors probably come into play.

Genes and Human Brain Size

Small differences between the genome of humans and non-human primates, especially chimpanzees, belie notable organ-specific differences in gene expression, especially in or affecting the brain. Nancy Minagh-Purvis (University of Pennsylvania) recently reported that a distinct mutation gives humans smaller *masseter* and *temporal* muscles than other primates. These muscles are on the side of the head and are used to operate the jaw, especially during mastication. She estimated that this mutation occurred about 2.4 million years ago and concluded that its main benefit would be to reduce the need for the big bony anchors on the ridgeline at the top of the primate skull, to which the enlarged temporal muscles anchor. Minagh-Purvis concluded that smaller jaw muscles in hominids might have reduced a significant constraint on expansion of the primate skull because of the reduced pressure created by these muscles on the skull. Once this pressure was removed, the skull could then accommodate an expanding brain. She suggests, however, that this expansion could have created an increased challenge for survival since the heavy food grinding capacity found in the hominid and non-human primate jaw would then have been lost.

Bruce Lahn and colleagues at the University of Chicago studied the *abnormal spindle-like microcephaly-associated gene,* which they speculate could have been involved in expansion of the human cerebral cortex. The normal form of this gene causes microcephaly, or impaired neuron development resulting in a smaller brain cortex. Lahn hypothesized that expression of an *abnormal* form of this microcephaly-associated gene preventing microcephaly may have been necessary for humans to evolve a fully developed cerebral cortex. By comparing the DNA sequence of the microcephaly-associated gene in humans, non-human primates and other mammals, Lahn's group was seeking evidence for whether this gene could have contributed towards explaining the minimal expansion of the cerebral cortex in non-human primates and its more marked expansion in humans.

They found that protein-changing mutations in this microcephaly gene increased as the species under study got closer to humans, but there were few such mutations in the cow, dog, sheep, cat, rat, mouse, or in more primitive primates. Such a clear trend in gene mutation across species is uncommon, so they speculated that an appropriate mutation resulting in expression of the abnormal form of this gene may have had a role in expansion of the human cerebral cortex because it permitted cortical neutron development in a way that seems not to be possible in non-human mammals possessing the normal form of this gene.

The N-Methyl-D-Aspartate Gene and Brain Function

The physical basis for learning is still unknown but seems more and more likely to involve long-term and simultaneous dialogue between neurons via well-established synaptic connections. Communication between two neurons can be strengthened by repeated use of synapses. This synaptic modifiability can be studied by employing *long term potentiation*, which occurs with brief, repeated stimulation of specific neural pathways, notably in the hippocampus. Long-term potentiation that increases the efficiency of transmission in hippocampal synapses can last from hours to days depending on the model. Several studies support the concept that long term potentiation is a necessary component of learning.

The N-methyl-D-aspartate (NMDA) receptor in the brain binds the amino acid, *glutamate,* which is a neurotransmitter controlling the initiation of long-term potentiation in the hippocampus. The NMDA receptor has a special combination of structural subunits that determines the amount of long-term potentiation that is possible. One such subunit, known as NR2B, is more abundant in young animals than in adults, and appears to be responsible for more efficient long-term potentiation and learning in younger compared to older animals.

At Princeton University, Ya-Ping Tang over-expressed the NR2B subunit of the NMDA receptor in adult mice. Three different types of learning were improved in the mice with the 'boosted' NMDA receptors – fear responses, associative learning and faster re-learning or override of a previously learned task. Effectively, the NMDA-boosted mice were smarter. Tang did not speculate about a possible role of the NMDA receptor in human brain evolution but this example shows how relatively simple gene changes affecting brain neurochemistry can have profound effects on brain function and learning capacity.

Protein Phosphorylation and Brain Function

Growth associated protein (GAP-43) is a very different protein from a neurotransmitter receptor like NMDA but it, too, is implicated in the relation between synaptic input, memory, and learning, so changes in its production could affect learning processes relevant to human evolution. Like with many other signaling processes, GAP-43's role in learning seems dependent on *phosphorylation,* or the addition of a phosphorus group to part of the protein.

If GAP-43 is involved in learning, in principle, modifying the GAP-43 gene to over-express the phosphorylated form of GAP-43 should enhance learning. Likewise, since phosphorylation is essential to GAP-43 activity, preventing phosphorylation of GAP-43 even in mice that overexpress this protein should prevent increased learning. Knocking out or preventing expression of the GAP-43 gene would be another approach in which learning should be impaired but this approach is not practical because mice do not survive deletion of this gene.

Aryeh Routtenberg at the University of Michigan showed that in several different learning trials, overexpression of phosphorylated GAP-43 in mice reduced the time needed to find the location of a hidden food sample, i.e. reduced errors in learning. This improvement was more apparent as the time delay between initial exposure to the test and subsequent re-testing was increased. Relearning, i.e. unlearning the original location and learning the new location of the food was also faster in the mice overexpressing GAP-43.

Because GAP-43 is present more at synapses than elsewhere in the brain, Routtenberg concluded that the two necessary precursors to true learning - long-term potentiation and synaptic connectivity - were both increased by overexpressing GAP-43. GAP-43 might also increase the release of certain relevant neurotransmitters at the time of the test, thereby inducing a form of *neurotransmitter priming*. This priming is in turn thought to improve long term potentiation by strengthening the signal coming back from the post-synaptic to the pre-synaptic cell.

Phenylketonuria and Other Forms of Mental Retardation

Molecular modification of genes clearly shows that learning can be experimentally manipulated in animal models. However, it is equally clear that several gene mutations affecting brain function cause mental retardation in humans. In fact, there are no known genes or gene mutations that specifically improve brain function in humans; they all seem to be detrimental and are known as *inborn errors of metabolism.* The disease - *phenylketonuria* - is a classical example. The term comes from the fact that a degradation production of an important amino acid, *phenylalanine*, is excreted in high amounts in untreated patients with this disease.

The metabolic defect in phenylketonuria involves inadequate conversion of phenylalanine to a second amino acid, *tyrosine.* Both phenylalanine and tyrosine are present in mother's milk and in most dietary sources of protein including milk formulas. These two amino acids are needed to build tissue proteins and other molecules including neurotransmitters.

When conversion of phenylalanine to tyrosine is impaired, incoming dietary phenylalanine builds up in tissues including the brain. Phenylalanine can be converted to some other metabolites but these other routes are not very efficient so, in the absence of the gene controlling conversion of phenylalanine to tyrosine, high levels of phenylalanine accumulate and become toxic to the brain. Until tests for this gene defect became routine, it was impossible to know that a child had this disease until it was already mentally retarded. Routine testing for the phenylketonuria gene at birth now gives sufficient warning that low phenylalanine diets can be given, resulting in more normal brain development.

Inborn errors are rare but often cause devastating diseases. They affect the synthesis or metabolism of many important molecules including cholesterol and glucose, and most minerals, vitamins, fatty acids, and amino acids. Many but not all of these errors cause mental retardation. None are known that enhance brain function. Nature seems to have left humans with no notably smart variants but has left many different (although still rare) defective variants. Perhaps the human brain is already operating at peak capacity. Gene mutations may be able to ramp up performance of certain neurotransmitters or signaling processes in lab animals but, metabolically or nutritionally, perhaps humans are already at or near the upper physiological limit of cognitive performance.

Genes, Schizophrenia and Brain Function

Until his recent death, David Horrobin of Laxdale Limited in Stirling, Scotland, was a prolific medical scientist interested in many aspects of the biology and clinical importance of polyunsaturated fatty acids. In his recent book, *The Madness of Adam and Eve*, he speculates that genes affecting metabolism of these fatty acids are involved in human intelligence and, more specifically, a darker side of human intelligence – *schizophrenia*. Schizophrenia tends to involve increased creativity, paranoia, intense religiousness and artistic achievement; in short, the essence of some of the defining characteristics of humanity but in extreme forms. Bearers of this disease are often incapacitated by it but positive (artistic and creative) elements of the disease are commonly seen in less-affected relatives.

Horrobin proposes that schizophrenia arose before human races evolved because it occurs in about 1% of the human population and appears to be evenly distributed between races. He has therefore reasoned that schizophrenia is a truly human condition that arose as an unfortunate product of the expanded intelligence that benefits the human species as a whole. In several studies, he has shown that schizophrenia involves slower uptake and faster release of polyunsaturated fatty acids in membranes than in healthy controls. He reasoned that this would reduce membrane polyunsaturates and adversely affect synapse development and remodeling during learning, thereby distorting the learning process and causing abnormal perception of what was learned.

Rather than focusing on single gene changes, Horrobin has speculated about coordinated changes in gene expression affecting several proteins involved in polyunsaturated fatty acid metabolism. These include proteins affecting body fat reserves, *desaturase* enzymes required for conversion of shorter chain to long chain polyunsaturates, *phospholipases* that cleave fatty acids from membrane phospholipids, and *fatty acid binding* and *fatty acid transport* proteins that would help get more fatty acids into the brain.

He notes that these multiple changes in fatty acid metabolism probably happened relatively recently and certainly after the emergence of *H. erectus*. Their impact would have been to promote the burst in creativity and early evidence of religious belief first associated with the Cro-Magnon.

Genes and Human Behaviour

The influence of genes on human behaviour is part of two disciplines - evolutionary psychology and behavioural genetics. Evolutionary psychologists propose that many behaviours specific to humans became hard-wired early in human evolution and that they should be traceable to human genes. Thus, by supposedly possessing brain structures that enable them to better detect cheating, preferred mates, or probable members of a good hunting team, etc., men are said to be genetically better prepared for food procurement. Hence, they rather than women evolved the culture of hunting.

However there are several problems with this reasoning. First, the 'fixing' of these behaviours requires that hominids occupy a specific habitat or environment that would have facilitated hunting. Would men still be responsible for food procurement if hunting weren't always possible? Second, human mate selection behaviours are also said to be hard-wired but culturally acceptable mate selection behaviour varies widely across human cultures and races. Thus, behavioural tendencies in one society (often Western/US) are naively assumed to apply universally and, furthermore, to be equivalent to evolutionary 'fitness'. Third, behaviour is a complex attribute presumably with input both from many genes and from developmental circumstances, so the hard wiring of behaviours in genes would take complex co-evolution of many genes.

Behavioural genetics is the other discipline addressing the role of genetics in human behaviour and cognitive advancement. It is built around the statistical comparison of identical and fraternal twins and has a long history of equating behavioural differences occurring in families with genetic differences. Twins are frequently used in studies of supposedly inherited characteristics like IQ, but on the basis of the totally inappropriate assumption that the environment of fraternal twins is the same as for identical twins. Furthermore, even if twins are separated at birth, their shared *prenatal* intrauterine environment is usually ignored.

In fact, if birth weight isn't the same, all bets are off anyway because factors affecting the rate of growth in utero (fetal programming) have a major impact on post-natal development and predisposition to disease (see Chapter 14). As Richard Lewontin has attempted to convey in several books, the problem is not only that no one really knows where shared genes and shared environments meet but also no one knows how a gene will behave in a new environment.

The Environment and the Heritability of Complex Traits

Heritability is a term that was originally applied to highly specific desirable (or undesirable) traits in animals or plants that were amenable to selective breeding. Heritability has evolved beyond its original intent and now more complex socially influenced characteristics such as high intelligence quotient (IQ) are said to be heritable, i.e. smart parents produce smart children.

However, the fact that IQ appears to show heritability establishes little about the role of genes in IQ because parents and their children share more than the same genes, they usually share the same habits and the same living environment. Heritability is a very local measure; it changes substantially with changes in genetic make-up or changes in the environment. Using Richard Lewontin's example, Roman mathematicians calculated slowly because they had a cumbersome numerical system not because they were unintelligent.

Indeed, if complex behaviours were truly heritable, it would be predicted that the high positive correlation between parents and offspring for both religious preference and political party was genetically governed, a proposal that would insult most people. Regardless of within family tendencies, both religion and voting preference are changeable showing that something in our cultural-social or nutritional environment modifies such preferences.

Even in fairly specific, non-controversial cases, heritability of a gene-linked disease indicates nothing about how easily the disease can be nutritionally managed. In some cases, the mutation is silent and no management is needed, such as in most of the more than 1,000 known mutations of the cystic fibrosis gene that have no detrimental effect on health. Clearly, in some cases, a mutation does cause symptomatic cystic fibrosis, which affects pancreatic, lung and intestinal function. Nevertheless, all known variants of cystic fibrosis are manageable to a greater or lesser extent with antibiotics and nutrition. As a result, life expectancy in cystic fibrosis is steadily increasing past thirty years old.

In other gene-linked diseases like phenylketonuria or acrodermatitis enteropathica, once the biological basis of the disease is understood, appropriate nutritional strategies can be devised and such individuals can lead normal lives including having children. Hence, although these inborn errors of metabolism are strictly speaking genetic in origin, the patient's nutritional environment plays a pivotal role in determining the eventual outcome of the disease.

Highly heritable behaviours are still modifiable by a change of environment. Indeed, empirically, prenatal and cultural environments must have a major influence because there are not enough human genes to accommodate existing gradations in human behaviour. The neurochemical basis for behaviour, i.e. interpreting what actions or reactions are appropriate, is genetically controlled, but specific genes alone cannot control individual behaviour or choices. Given the relatively small number of genes thus far known to be associated with learning and behaviour, consensus is emerging that most genes involved in human learning and behaviour are *pleiotropic*, i.e. must be involved in more than one process.

The advantage of such a situation is that a mutation improving one function is more likely to be selected for because it influences several other functions as well. More importantly, the plasticity of behaviour leads to contrasting benefits of the same behaviour, e.g. impulsivity, which can be advantageous in a fight-or-flight situation but disadvantageous for mate selection. This plasticity doesn't rule out a gene effect; it just shows a lack of predictive ability of a supposed gene effect once the environment changes.

Even when the role of the environment in a complex condition like heart disease is acknowledged, it is common to try to estimate the relative roles of genes and the environment in the risk of that disease. However, it is fallacious to set percent contributions of genes and the environment to a particular attribute and, without experimental testing, to try to predict the amount these proportions will change in a different environment. Heritability does not confer fixity. Gene effects are not a fixed property but vary according to the environment. Differences amongst humans, i.e. in behaviour, intelligence or in disease risk, are an additive consequence of both genetic and environmental differences; the two are not independent causal pathways.

Richard Lewontin makes the case (is) that genes have three functions: (1) To specify the exact code for the composition of proteins, (2) to specify the start and end of protein production; thus, even though they all have the same genetic code, cells of different tissues can be different and developmental maturation can occur, and (3) to be the pattern to make further copies of themselves.

Genes are commonly said to make proteins and to be self-replicating but Lewontin points out that in the absence of the appropriate cellular machinery and raw materials, genes make nothing. The genetic code is called a blueprint but it is a blueprint that is not functional in the absence of the relevant raw materials, i.e. amino acids to build the proteins, other nutrients that are substrates for lipids and most complex molecules, etc. The blueprint cannot work without the machinery or the raw materials and vice versa; neither is more important than the other.

***

Perhaps in the not too distant future there will be ways in which studies like Routtenberg's involving overexpression of a certain mouse gene that increases 'smartness' can be done in the same mice in which Bruce Lahn's microcephaly gene is simultaneously knocked-out. An approach of this nature might allow us to see whether combined mutations could have a cumulative positive effect on learning in mice. It would probably be easier to see a human-like impact of such changes if the gene modification were done in primates but this type of experiment still seems a long way off.

Are there in fact differences in GAP-43 expression or any of these other genes between humans and non-human primates? Where such gene differences have been identified, do the proposed explanations exclude other equally plausible interpretations?

In the case of the gene discovered by Minagh-Purvis, could it be that a better quality diet permitted or even promoted loss of the jaw muscles whose shrinkage in humans she attributes to a gene mutation? The diet effect might or might not be dependent on certain nutrients affecting the size of jaw muscles. Perhaps there was no scramble for survival at all, but rather the combination of an improved diet and loss of the gene for large temporal and masseter muscles was advantageous to human brain expansion.

Underlying experiments on the molecular or genetic basis of brain function and evolution is the question of whether gene mutations could lead towards the human capacity for intelligence only in an already human-like animal, i.e. in a primate at least as cognitively advanced as a chimpanzee. Making a smart mouse is an impressive accomplishment and shows that highly specific changes in genes can have a marked influence on that animal's performance on selected tasks. Certainly, if translatable to humans, even a 10% improvement in learning ability would be a powerful change, but would it have adverse effects?

The positive attributes of the mouse learning experiments are widely publicized but would such an improvement apply equally in a more developed (primate) brain? Would the same overexpression of GAP-43 make humans more creative or actually more devious, dangerous or likely to harm others? Mice aren't able to be sinister so it may difficult to see or directly monitor the negative potential of such experiments when done in species smarter than a mouse. As Lewontin has remarked, if the true basis of human intelligence is to be understood, the strong desire to oversimplify its fundamental complexity needs to be contained.

Chapter 10

Bringing the Environment
and Diet Into Play

Empirically, we know from everyday experience that environmental stimulation has an important effect on human behaviour and function, whether the stimulus involves one's diet, one's family activities, or one's broader social environment. Two broad categories of environmental stimuli affect brain function – culture and nutrition. Culture involves people interacting with people thereby communicating skills and acquired knowledge that affect individuals or the group as a whole. Nutrition represents the constituents of foods, some of which are useful as energy substrates and some of which are substances essential for the structure of the brain.

Charles Darwin thought that cultural factors were a major stimulus for human brain evolution. Many prominent biologists support the idea that the modern human brain arose primarily because early humans developed more advanced culture than hominids or primates. Ralph Holloway and Harry Jerison propose that culture only came into play once language, a key antecedent of culture, emerged.

Language was certainly a decisive factor in developing a cooperative social milieu and, in some uncertain way, was dependent on expansion of pre-human group size. The connection between group size, language, culture, and brain stimulation is a safe bet but it leaves two important questions unanswered: What stimulated larger group size and how does brain stimulation lead to brain expansion and evolution?

Gene-Culture Co-Evolution

In *Promethean Fire: Reflections on the Origin of Mind*, Charles Lumsden and Edward Wilson propose that the human mind evolved as a product of *gene-culture co-evolution*. Lumsden and Wilson's concept of gene-culture co-evolution involves a genetic influence promoting culture via changes in what they consider to be well-established behaviours all humans have in common, regardless of culture or race, including: outbreeding favoured over inbreeding, mothers bonding with infants, certain colour preferences, shape preferences during early development, consistent facial expressions for the main moods and emotions, the capacity to discriminate the taste of different sugars, acid, salty or bitter, anxiety at an early age in the presence of strangers, and certain phobias.

According to Lumsden and Wilson, these universal human behavioural features are all independent of culture, and all follow *epigenetic rules*, i.e. rules that create recognizable, consistent stages in mental development that lead to an *innate preference* for one mental operation over another. They suggest that these rules form an underlying structure that jointly guides human cultural and mental development and ultimately supports human survival.

Culture influences genes because ways in which the mind develops or is likely to develop increase the odds of certain reproductively advantageous outcomes or behaviours. This leads to more efficient reproduction, thus further increasing those behaviours in the population and influencing the gene pool. Humans are therefore a product of gene-culture interaction, which predicts rapid evolution of the human brain and transmission of culture and cultural diversity that varies according to the behaviour in question.

Like Lumsden and Wilson, in *The Blank Slate: The Modern Denial of Human Nature,* Steven Pinker claims humans have a gene or genes to fear spiders and snakes. If humans are born hard-wired with such phobias that aid survival, this is important evidence that at least some of our most fundamental behaviours are laid out in our genes, ready to function on first exposure to the relevant stimulus.

However, in *Guns, Germs and Steel,* Jared Diamond argues that human fear of spiders is a learned response. When human infants see older children or adults grimace at the sight of a spider, they quickly mimic this response just as they mimic many other responses or behaviours they see. Diamond notes that there is a high abundance of both spiders and snakes in New Guinea but children there show no fear of spiders and, indeed, eat some of them! Pinker acknowledges that apparent non-fear of snakes exists in some aboriginal peoples but claims the innate fear is overcome by practice and sufficient exposure.

Clearly it is important to primate and human survival in woodlands or jungles to have a fear of dangerous predators, including snakes and spiders. The issue is whether fears are imprinted or whether the fear is already there prenatally, pre-packaged and ready to operate at first exposure. This example is fundamental to the debate over nature versus nurture – how much of human behaviour is learned versus innate.

A mother's face (and smell) also gets imprinted so I would argue that *imprintability*, i.e. the ability of infants to learn certain responses very rapidly, is what is hard-wired in our genes. However, a high capacity for imprintability does not mean that a repertoire of specific imprints is pre-packaged in human genes.

There are many well-documented examples of young animals imprinting on genetically-unrelated or inanimate objects as their parents. These objects have no intrinsic survival value to the infant animal but they are imprinted as supposed parents because virtually any object seen often enough very early in life is most likely to be a parent bringing food, warmth or comfort. Ability to recognize the food bearer is imprintable and hard-wired but the form or even the species of the food bearer is flexible.

Trying to establish which behaviours are specific to humans and which aren't misses the point. The underlying point is that, compared to primates, humans have an expanded range of imprintable behaviours. As Richard Lewontin has elegantly argued, how specific behaviours became part of the overall human behavioural repertoire isn't the issue; the issue is how the range and depth of the human behavioural repertoire evolved. In fact, it is the loss of specific, rigid behaviours and the expansion of a wider range of flexibility in what can be learned that is uniquely characteristic of humans.

Genes appear to be responsible for the neurochemical and molecular machinery making rapid and flexible imprintability of behaviours possible. That rapid imprintability makes some apparently universal human behaviours appear innate. It also makes other behaviours that will emerge later in development more or less likely. The learned behavioural repertoire varies according to the habits and behaviours of different human groups and forms the basis for culture.

What remains to be found is a mechanism by which the possible behavioural repertoire of pre-humans expanded during human evolution. Culture is a product of that repertoire and can shape it but culture didn't create the expanded template for human behaviour. Specific genes create the blueprint for specific biochemical reactions but human behaviour isn't about specific reactions, it is about the evolution of expanded flexibility (and vulnerability) in learning as well as behavioural plasticity that is not limiting or detrimental to survival.

Extraordinary Circumstances

We take it for granted that the human mind is distinct because of powerful long-term memory, symbolic thinking and language. Lumsden and Wilson point out that, actually, the human mind is also remarkably inefficient in some ways. They note that humans have poor ability to recall rote information, to judge the merits of other people, or to estimate risk or plan strategy. Humans also make excessive use of stereotypes in planning strategy, and also tend to equate events that have low probability and trivial consequence with events that have low probability but important consequence. Therefore we underestimate the effect of catastrophe even when the identical events are repeated and experienced by sequential generations, i.e. warfare, floods, windstorms, drought, etc.

Unlike whales or songbirds, which can faithfully repeat long complicated sound sequences in their songs, humans have poor *kinesthetic memory* – the ability to exactly replicate sound or movement sequences. Only *idiots savant* are capable of repeating the numerical value of 'pi' to many decimal places. Humans also have poor spatial and temporal memory compared even to simple-brained insects like honeybees.

Possession of these capabilities at a higher functional level, i.e. ability to repeat species-specific sound sequences or to recall events in time and space would seem to have been advantageous to human survival. As suggested in Chapter 4, the cognitively advanced human brain seems not to be especially well developed around attributes that would seem to favor survival; some other much simpler animals are able to make these survival 'decisions' better than are humans.

How is it that humans have such an advanced behavioural repertoire yet they are decidedly inefficient in survival behaviours or skills? Lumsden and Wilson try to address this paradox by arguing that while it would be advantageous to individual humans to have improved kinesthetic memory or to be better able to judge the merits of other humans, the need for such skills appears to be lower when living in large groups, i.e. in society. They therefore argue that some aspect of the environment actually perpetuates a certain degree of 'inefficiency' in human mental processing, perhaps because those mental skills are not crucial to the group's survival, only to an individual on its own.

Lumsden and Wilson are unimpressed with the classical explanations, including Darwin's, that war or sexual competition promoted human brain evolution. They also pan the idea that extreme temperature effects - both heat stress from hunting on savannahs and the stress of a cold environment - might contribute to brain expansion, even though survival is dependent on solutions to these problems.

They speculate that *'some extraordinary set of circumstances – the prime movers of the origin of mind – must have existed to bring hominids across the Rubicon and into the irreversible march of cultural evolution.'* Extraordinary circumstances would indeed have been needed to help perpetuate such a remarkable combination of human mental inefficiencies that all bear more or less on chance of survival, i.e. poor memory for sound, place and time, or poor judgement of character or risk of catastrophe, while at the same time facilitating development of skills with no survival immediate value, including art, music, and religion.

As a solution, they turn to *H. habilis* whom they claim was not skilled in any particular way but was nevertheless *'pre-adapted for intelligence'* by already having a large brain. They speculate that *H. habilis* had to evolve better intelligence to compete, which *'nudged the human line forward until gene-culture co-evolution ignited and became a self-sustaining reaction'.* This competition ignited cultural innovations, which, in turn, acted as a new class of mutations – *memes* - that *'accelerated evolution and pushed the species forward to its present genetic position'.* Memes are bits of culture that maximize their own replication, somewhat like Richard Dawkins' selfish genes.

The problem is that this argument is circular – *H. habilis* was just sufficiently cognitively advanced to respond to a timely need to compete. How did *H. habilis* get sufficiently advanced and what exactly was the need for competition – for food, sex, or territory? Many humans are undeniably territorial today but the catalyst to get *H. habilis* sufficiently advanced to move on to Lumsden and Wilson's cultural Rubicon remains a mystery. How, indeed, did *H. habilis* become competitive despite spatial and temporal forgetfulness, inability to judge the merits of others or of risks in general?

Like Lumsden and Wilson, Paul Ehrlich and Marcus Feldman at Stanford University propose that human behaviour is a product of a diverse cluster of non-genetic information they call culture, but Ehrlich and Feldman couple culture to chance and natural selection. For them, chance constitutes short-term random mutations in genes whereas natural selection represents long-term change in genes supported by successful reproduction. Given that vision is the principal sensory means by which humans interact with the environment, they suggest that human perception of the world and our opportunity to evolve a culture is primarily based on visual cues. Ehrlich and Feldman suggest that evolution of more sophisticated *interpretation* of visual stimuli (in the association areas of the cerebral cortex) had a role in human behavioural evolution.

Separate External and Internal Environments

In a simplified form, Darwin's message was that evolution involves organisms that adapt or die. Organisms that successfully adapt are able to reproduce and leave successful offspring. Darwin's central point was that the environment is separate from the organism; external and internal environments are separate. The environment throws up the problems and the individual attempts to solve them. Various solutions appear more or less at random with individuals bearing the most correct ones being preserved.

To Richard Lewontin, too much of modern biology promotes the idea that nature has dealt a set of cards that cannot be altered so it must be our internal environment, i.e. our innate differences or our genes, that makes some humans fit and others unfit for nature's fickleness. This is the foundation of *biological determinism*, which, as applied to humans, has three key attributes - (1) individual humans differ in fundamental abilities because of innate differences, (2) these differences are inherited, and (3) human nature guarantees the formation of hierarchical society.

Genetic determinism accepts that an organism's features are a consequence of genes plus an effect of the environment but claims that the environment simply extracts or magnifies the genetically-determined differences. In contrast to the premise of both biological and genetic determinism that innate capacities are fixed, Lewontin notes that, in fact, capacities vary a great deal according to the environment: In one environment, one capacity may be superior, but another environment can make that same capacity inferior.

The tasks that demonstrate the fallibility of the fixity of innate capacity do not need to differ that much. One strain of mice may be better at one task than a second strain but worse at a different task, yet their genes haven't changed when the learning task changes. Hence, from the genes alone, it cannot be predicted what differences in learning capacity will be exposed on different tests, i.e. in different environments.

In lab animals with long histories of separate breeding that give rise to identical genes, differences truly attributable to genes are relatively easy to distinguish but only in identical environments. This type of experiment is much harder to do in people who share not only genes but also family history, culture and social structure. Studies of identical twins raised apart should give best evidence of role of genes in a trait since they have identical genes but are located in different environments.

Developmental 'noise' is the final wrench in the gears of the simplistic gene versus environment, nature versus nurture, debate. Developmental noise is the persistent variation that still occurs even when genes and the environment are identical. Lewontin cites the example of different numbers of hairs on one side compared to the other side of the belly of the housefly. The fly's environment is the same as are the genes controlling its belly size and shape but the number of hairs is not. Genetically identical worms raised in the same container in the same lab have different insulin sensitivity and different longevity. Developmental noise is the uncontrollable variable that gets between the environment and genes. The reasons for this variability are unclear. The debate is really not between nature and nurture but is really a poker game between nature, nurture and noise, with noise being dealt the occasional wild card that the other two never get.

Individuals, Society and Environment

Lewontin laments the fact that the search for humanness in our genes engenders a reductionism that misses the point that human social organization exceeds the apparent limitations of the individual. He uses the example of human flight, which, unaided by aeronautic technology, is impossible for an individual or even for a group of individuals. Yet powered human flight happened. Human flight was a consequence of human social organization - organization of science, education, discovery of gasoline, monetary systems, air traffic control, electronic communication, etc. Nevertheless, society does not fly. Individuals do, but only as a consequence of social organization. As Lumsden and Wilson suggest, individual limitations do not necessarily limit individuals in a society.

Society creates a dynamic that cannot be understood by looking at the individual removed from its society (environment). This not only applies to technological achievement but also to human susceptibility to certain diseases, which do not affect the individual except when they are massed in (and, in this context, disadvantaged by) society. In both cases, the nature of what is human is a product of the environment, an environment created by individuals.

This is not to say that human biology is irrelevant to human behaviour, competencies or social organization. Lewontin cites body size (5-6 feet tall) as a critical feature that gave humans the physical strength to acquire the early technologies of smelting, mining, fire control, etc., all of which were necessary for modern technology. Although intelligence is more than a product of brain size, it still probably takes a minimum brain size to have enough neuronal complexity for speech. For their size, ants are very strong, clever and socially organized but they will never get to the moon (except, perhaps, as opportunistic passengers).

Genes certainly helped make humans physically big enough and provided the blueprint for larger brains with advanced memory and cognitive function, yet there are not enough genes to account for human consciousness, which is the single most important feature creating the human environment. From history, we can see what human consciousness has accomplished but if history teaches us anything it is that we cannot know what humans are capable (or incapable) of in the future.

Adaptation and Exaptation

Adaptation is another mechanism by which new features (including behaviour) are said to evolve. Every aspect of an organism's morphology, physiology and behaviour can be viewed as adaptation to challenges faced by the organism. Adaptationists propose that a feature is present to address a preexisting challenge to survival. This makes it a somewhat reactive approach to evolution, i.e. that the human brain is designed in response to environmental demands, such as the need for tools to hunt.

For instance, the adaptationist explanation for bipedalism in humans is that it is a more efficient (less energy consuming) means of locomotion for humans than the quadrupedalism of chimpanzees. Bipedalism also freed the hands, which were needed for tool use. Tool use was needed for hunting. Hunting was needed to fuel brain expansion, which was adaptive for culture.

Wait a minute – bipedalism may be more efficient today than quadrupedalism (although even that is still debated). However, early bipedalism could hardly have been more efficient in the beginning when hominids were tottering around trying to get their balance. This rationale requires the first hominids to adapt to a more primitive form of locomotion (early bipedalism) than they previously had so that they could eventually become more efficient bipedal walkers (100,000 years later, maybe more?).

If energy efficiency were the driving force of bipedalism, why wouldn't chimpanzees also adapt to it? Natural selection doesn't anticipate eventual improvements. To be selected for, each step has to have a net positive benefit to those individuals in the species that display the modification as it is occurring rather than for whatever value it might have in later generations.

Ian Tattersall of the American Museum of Natural History in New York is reluctant to specify what he thinks might have been the stimulus for human brain evolution. In *Becoming Human*, he takes a minority and somewhat courageous view that once the larger human brain had evolved it became *exapted* for improvements in hunting, language, behavioral refinement, and other skills. Hence the brain expanded (somehow), and then hominids began to take advantage of the brain's improved capabilities, i.e. they exapted it to their behavioural advantage.

Reflecting on the classic explanations for human brain evolution, i.e. meat eating by scavenging or hunting, Tattersall is concerned that although scavenging has a rather undignified reputation, hunting falsely embellishes the human self-image. He feels that the idea that hunting was the catalyst for human brain expansion has arisen from a *'deeply flawed interpretation of our past in terms of our present'*. He proposes that *'something else'* promoted human brain expansion, which then created the opportunity for language and other cooperative social skills. Once established, these developments were in effect self-propagating and led to culture, hunting, technological achievement, etc.

I agree with Tattersall that *'something else'* promoted human brain expansion, from which language, culture, art, religion and technological advances took advantage, i.e. became exapted. Tattersall comes closer than anyone I have read to suggesting that human brain expansion was an undirected, passive process; it had no purpose and did not occur to increase chances of survival. Everyone agrees that natural selection is supposed to work that way but few explanations for human brain evolution actually adhere to this demanding rule.

Kevin Laland and Gillian Brown at Cambridge University propose that human behaviours are *'by-products of our extraordinary adaptability rather than being adaptations'*. To me, this is the same as Tattersall's idea that the human behavioural repertoire is an *exaptation*, i.e. taking advantage of a modified feature (the enlarging brain) for a new use once it is in place. In some cases, the modified feature may have evolved for other reasons. In contrast to adaptation, adaptability or exaptation doesn't lock in any specific behaviours, circumstances or outcomes; anything could happen, good or bad.

Adaptability or exaptation as a part of human brain evolution is an important yet rare idea. I would equate these terms with *plasticity*. Plasticity during development is tolerable as long as one is not too vulnerable to negative outcomes. Plasticity and vulnerability are opposite faces of the same coin – human brain development has high plasticity and high vulnerability. Chimpanzees have somewhat lower neural plasticity but they also have somewhat lower vulnerability. Goldfish have very low neural plasticity but they have very low vulnerability (see Table 3.1). Increased vulnerability generally implies decreased chance of reproductive success and survival, so for plasticity to permeate neural development, vulnerability has to be keep to a minimum – it has to be masked by developmental stability.

Neural plasticity and low vulnerability in humans must be a product of high vulnerability being masked by environmental, developmental and nutritional stability. High vulnerability cushioned by environmental stability is a condition that no other mammals besides humans, even primates, ever experienced. Goldfish inhabit an environment that is notoriously fickle with extremes of oxygen and temperature so they were forcibly excluded from the high vulnerability option and had to accept the low vulnerability - low neural performance option.

Humans, and hominids before them, adapted to the high vulnerability – high performance option because they increasingly inhabited an environment that was stable and rich in both brain energy substrates and brain selective nutrients. Essentially, this left them less and less exposed to their underlying developmental vulnerability. Neanderthals appear to have had their neurodevelopmental vulnerability reexposed by less stable environmental conditions (possibly worsening diet; see Chapter 13) that cut into their adaptability and reduced their cognitive performance relative to early humans.

Diets – A Key Attribute of the Environment

Accepting Tattersall's view that hominids exapted their evolving brains to uses that were eventually beneficial, and Lewontin's point that different environments modify the phenotypes of identical genomes, it seems reasonable to propose that human brain evolution required a significantly different environment (different *'conditions of life'*) from that inhabited by other primates. Still, there is little useful evidence in the fossil record to guide the search for the essential features of this seminal environmental difference.

The main clue in fossils is that hominids had smaller jaws and teeth than do the great apes. Smaller jaws and teeth suggest less need for strong jaw muscles and large tooth surfaces to grind fibrous plant food that has low energy density. Hominid fossil specimens with larger crania generally have smaller jaws and teeth, especially in the last two million years. Hence, early humans not only have larger brains than did hominids but hominids gradually acquired a better quality diet than was consumed by other primates.

Over many years, anthropologists have taken a broad look at the impact of the diet in human evolution. For instance, in *People of the Lake,* Leakey and Lewin suggest that the turning point in human history came when people started to live in villages and had food that could be stored because it was available in excess of their immediate requirements. Once a long term abundant food supply becomes available in a single place, they suggest that the constraints on population growth imposed by a nomadic, hunter-gatherer lifestyle are lifted. Birth rate then increases, and the population starts to grow and accumulate possessions.

Leakey and Lewin note that contrary to common belief, the key feature of successful village life is not necessarily agriculture but simply a plentiful and reliable supply of a staple food in a central accessible location. It might be corn or wheat but it could equally be fish. Hunter-gatherers may well have a plentiful supply of food but they have to move around to exploit it, thereby preventing a material economy or fixed physical location for developing more complex social structure. Leakey and Lewin are not specific about which hominids their concept applies to, nor where they might have existed.

Twenty years ago, Robert Martin added the issues of energy intake and brain energy needs to the melting pot of ideas on dietary influences on human brain evolution. Others such as Tim Clutton-Brock at Cambridge University had already started to deal with the primate brain's energy requirements in general terms. By suggesting that the brain's expensive metabolism would have limited brain expansion unless dietary energy could keep pace with the increased energy demand, Martin seemed to nail down more clearly than anyone at the time the necessity of a better quality diet.

Amongst those who ran with the 'better quality diet' concept in relation to human brain evolution, Bill Leonard of Northwestern University near Chicago stands out. In collaboration with various colleagues, he has published numerous articles describing the need for a higher quality diet to support the energy demands of a larger brain. In his recent work, he reports semi-standardized measures of diet quality that take into account the contributions of different structural parts of edible plant material. Thus, the bark, leaves, and stems offer the lowest diet quality. Reproductive plant material, e.g. fruits and flowers, are of higher quality, but intake of animal parts provides the highest energy and nutrient quality to the primate or human diet.

Leonard notes that large-bodied primates generally have a combination of lower metabolic rate, lower quality diets and proportionally smaller brains. Small-bodied primates have a higher metabolic rate and larger brains relative to body weight. He notes that humans are exceptional in having both a better quality diet and bigger brain size than predicted from body size. Leonard provides well-reasoned and well-documented support for these observations and quite reasonably concludes that the possessing a bigger brain '*necessitates*' a better quality diet; pre-human hominids had bigger brains so they had to find a better quality diet.

Like many before him, Leonard adheres to the Savannah Theory, eg. that a drier climate in east Africa up until about two million years ago would have reduced woodlands and promoted emergence of savannahs. In effect, a harsh climate forced the newly bipedal Australopithecines to adapt to a grassland habitat for survival. Savannah grasses provided lower energy density than did the fruit-bearing trees, bushes and other plants of the disappearing woodlands. However, the savannahs encouraged the establishment and proliferation of grazing animals – antelope, gazelle, etc., which became part of the pre-human hominid diet.

After a long period of adaptation during which Australopithecine brain size changed little, the tool using capacity of *H. habilis* began to emerge. This newfound skill was put to the test by occasional and then more regular recovery of meat from carcasses remaining from lion kills. Sharpened cutters and scrapers made of stone would have been used to remove and cut meat. They also served to remove and clean the hides that could then be used for clothing and shelter. Increasing intake of meat more than made up for the lower energy yield from savannah grasses that would have been the only real alternative. Meat intake increased diet quality and met the increasing energy needs of the larger body and brain as *H. erectus* evolved from *H. habilis*.

Hominid fossils from this period clearly show changes in jaw shape and structure as well as in dentition that are consistent with lower intake of fibrous plant material typical of that consumed by present day non-human primates. Leonard and colleagues also point to fossil evidence suggesting that as Australopithecines evolved into *Homo*, total musculature decreased compared to other primates and, indeed, compared to all other comparably sized mammals. Thus, improved diet quality contributed to a net energy surplus in early *Homo*, especially earlier *H. erectus*, in whom brain size was markedly increasing. This energy surplus was used to fuel three metabolic requirements - larger body size in female *H. erectus* than in earlier female hominids, the larger brain, and fat deposition, especially in infants.

In work separate from but relevant to Leonard's, Leslie Aiello and Peter Wheeler have noted an additional energy saving process as hominids started eating meat - the gut decreased in size as the consumption of vegetation was reduced. Since the gut is a significant component of body weight and is much larger in vegetarian primates than carnivores, if the gut had lower energy needs as *H. habilis* or *H. erectus* ate more meat, that could tip the net whole body energy balance further into surplus.

What Exactly was the Better Quality Diet?

Over the past twenty years, Boyd Eaton and colleagues at Emory University in Atlanta have developed the case that the diet consumed by present day aborigines who maintain their traditional hunter-gatherer lifestyle is a good model for 'paleolithic nutrition' in hominids and humans. Eaton has calculated a range of possible nutrient intakes from the wide variety of plants and meat consumed by various groups of extant hunter-gatherers. These data have been compared to present-day food and nutrient intakes of more sedentary populations in industrialized countries. They show that the intake of saturated fat and sodium is higher but intake of slowly absorbed carbohydrates, protein, dietary fiber, polyunsaturated fatty acids, and vitamin C is lower in the more affluent, 'non-traditional' groups. Eaton describes meat as a source of the two omega 3 polyunsaturates – eicosapentaenoic acid and docosahexaenoic acid. (Actually, meat from wild animals is a very poor source of these important fatty acids.)

Eaton proposes the that lower incidence of chronic degenerative 'western' diseases (cancer, diabetes, obesity, hypertension, atherosclerosis) in aboriginal groups still adhering to their traditional lifestyles is due in large measure to their diet, which provides less total fat, saturated fat, and sodium but higher intake of slowly absorbed carbohydrates, fiber, vitamins and polyunsaturated fatty acids. Present day advice from a wide range of government and public health agencies is consistent with Eaton's assessment and lays much of the blame for chronic 'killer' diseases at the feet of the current 'western' diet, which is widely felt to contain too much refined sugar and fat, and too little fruit, vegetables, fiber, and slowly absorbed carbohydrate.

Eaton has reasonably assumed that food procurement patterns of present day traditional hunter-gatherers are a proxy for hominid food intake patterns. Given that the database from which the average hunter-gatherer diet composition was based, Eaton comes to the unsurprising conclusion that hominids had a lower intake of fat and a higher intake of fiber and vitamins than at present. Noting that later hominids were at least as tall if not taller than humans today and that many specimens had heavier bone structure, Eaton suggests that some hominids probably had higher meat intake, and much more strenuous lifestyles than humans have at present. Despite a wide-ranging discussion in several published reviews of the relevance of nutrition to human chronic disease incidence, Eaton discusses neither human brain evolution nor the extremely high fat intake of the Arctic Inuit.

Human Tolerance of Variability in Diet Composition

It needs to be clear that wider extremes of diet than those proposed by Leonard, Eaton or others can still easily meet known human nutrient requirements and can also be physically tolerated for long periods by humans today. Both lower chronic disease risk and improved physical performance, especially by athletes, are widely thought to occur when carbohydrate is the principal component of the diet. With physical performance, the logic is that since glucose supports energy expenditure of muscle, and glycogen is the storage form of glucose, therefore high dietary carbohydrate helps maintain maximal tissue glycogen levels and optimize physical performance.

In a 2004 article in *Nutrition and Metabolism*, Stephen Phinney succinctly and explicitly lays out the case that it ain't necessarily so. Phinney is a physician scientist particularly interested in the relation between diet, health and physical performance. He, too, adhered to the concept that moderately high carbohydrate intake was essential for optimal athletic performance, a concept supported by his own controlled metabolic experiments showing that endurance of both athletes and soldiers in the Canadian Arctic was reduced when very low carbohydrate diets were consumed.

However, Phinney was puzzled by two nagging but well-established examples in which humans could maintain long periods of high endurance under adverse physical conditions while eating almost no carbohydrate. The first example is of the Arctic Inuit (previously known as Eskimos) who endure extreme cold for long periods yet eat almost no carbohydrate except for a brief period in late summer when berries can sometimes be collected. The impressive tolerance of low carbohydrate intake together with unimpaired physical performance by the Inuit could arguably be based on a form of genetic or metabolic adaptation occurring over the 5,000 years or so that the Inuit have nomadically occupied the Arctic.

Adaptation cannot account for the other example, which is of the Europeans who explored the Arctic over the past 150 or more years. Their diaries are the principal source of information about what they ate, the length of their expeditions, and the distances they covered. In that sense, their reports are unproven and anecdotal. Nevertheless, as Phinney notes, over the course of an Arctic winter, a dog sled team can only carry so much food after which one relies entirely on what can be snared, harpooned, hunted or hooked. Reindeer were and still are a key source of food in the Arctic. These explorers of European origin describe the expected fatigue of switching to extremely low carbohydrates once their original rations ran out. However, they go on to recount how that phase passes as the body adapts to prolonged though moderate ketosis.

Outsiders never seriously questioned the duration of these expeditions. Still, the possibility that they were completed while subsisting exclusively on meat and fat was scoffed at. Hence, to resolve the ongoing controversy of whether humans could survive while fat was the principal component of the diet, one of the explorers agreed to recreate his Arctic diet under highly controlled conditions in a metabolic research setting in New York.

Constant observation guaranteed that over a twelve month supervised experiment, a diet comprising only meat and fat was consumed. This regimen resulted in sustained ketosis yet absence of nutrient deficiencies, especially the much-anticipated scurvy caused by vitamin C deficiency. Nutrient deficiencies were not only absent but, in fact, the two subjects maintained good health and physical performance as they had claimed for their period while isolated in the Arctic.

The two key elements of long term adaptation to ketosis in humans are, first, that health and physical performance need not be impaired by a low carbohydrate (*ketogenic*) diet and, second, that the adaptation period takes about three to four weeks before the previous level of physical performance is reacquired. Hence, short term studies of physical performance lasting less than two weeks will indeed find lower performance while on very low carbohydrate intake but that situation changes when the adaptation to very low carbohydrate intake extends for longer periods.

Two features of a very high fat diet are crucial to maintaining physical performance and preventing fatigue. These are the maintenance of mineral (sodium and potassium) balance and the maintenance of an optimal range of protein intake – neither too low nor too high. The Inuit and Caucasians who have lived with them have adapted to these limiting features of sustaining ketosis.

Like Phinney, Loren Cordain of Colorado State University has similarly cautioned that meat intake exceeding about 100 grams per day causes metabolic problems and cannot be sustained by humans. Hence, the key element in the Inuit diet is a very high fat intake. This leads to the highest intake in the world of long chain omega 3 polyunsaturates but is also the highest intake by any aborigines of total fat, regardless of composition.

I agree with the interpretation of Eaton and others regarding the range of diet composition that supports optimal health in present day, sedentary peoples. I accept that many traditional hunter-gatherers in warmer climates than the Arctic eat low amounts of fat and high amounts of fiber, fruit and vegetables. Indeed, some appear to be nearly as vegetarian as other primates. Like Eaton, I dislike the pervasive presence of refined sugar, adulterated fats, and innumerable additives in the 'modern' diet.

Disregarding the inconveniences and short-comings of the refined, modern day, western diet, I support Phinney's case that an extremely high fat intake with modest intake of protein and mineral balance is compatible with healthy human subsistence and physical performance. Humans can in principle survive on a wide range of diets in which fat can represent from 10 to 90% of energy intake.

However, I acknowledge that when offered a choice, humans do not typically choose diets containing 90% fat nor 90% carbohydrate nor, indeed, do they chose 90% protein for that matter. My point is that once minimum nutrient requirements are met, a very wide range of dietary fat intake is entirely compatible with human existence. We should therefore be cautious of developing an excessively specific or narrow range of carbohydrate, protein or fat composition of a single idealized paleolithic diet that minimizes chronic degenerative disease risk in present day humans.

Chapter 11

The Shore-Based Scenario: Why Survival Misses the Point

How did pre-human hominids obtain a better quality diet than was available to other primates? The fossil record clearly shows that around two million years ago hominid brains were starting to expand noticeably. By this time, hominids had learned how to make sharp stone tools and were using them to hunt animals for meat. Meat is a good source of energy and nutrients and early humans are widely thought of as having been hunter-gatherer-scavengers. Much later on, fossil evidence shows that some hominids, especially Neanderthals, ate a lot of meat. Hunting would certainly have helped meet the increasing energy requirement as the brain got larger.

However, there is a problem with hunting giving rise to big human brains. In fact there are several problems: First, in order to conceive of using sharpened stones as weapons and then to use them to hunt effectively, one would already need some degree of cognitive improvement before the brain expanded. How would that have been achieved if it was the proceeds of hunting and scavenging that fuelled the initial brain expansion? Second, one would need to demonstrate that the meat acquired by hunting was available to infants and children, because they are the ones with the rapidly developing, vulnerable brains. Third, although present-day humans possess big brains and have the potential to be very intelligent, few humans are skilled enough to hunt with simple weapons.

Successful hunting takes time and practice so, even when stimulated by hunger, few humans without considerable experience can successfully trap or kill an animal for food. Hunting is a learned skill but not under the duress of imminent hunger or starvation. The ability of early hominids to make the tools and learn the skills of hunting could have evolved no faster than their brains evolved and certainly could not have preceded brain enlargement. Initially, hominid brains weren't any bigger than those of chimpanzees today. Early attempts at hunting would therefore have been clumsy and dangerous, and would not have met the energy needs of many people, let alone those with rapidly growing brains such as babies and children.

Chimpanzees have smaller brains than humans and are not especially frequent hunters. Nevertheless, in small groups, they are very good opportunistic hunters. Three important points emerge from this example: First, primates with smaller brains than humans can still be efficient, cooperative hunters. Second, opportunistic hunting doesn't necessarily give rise to a larger brain or to dependence on hunting. Third, chimpanzees are not very good at sharing a killed monkey or other game, so early hominid hunters cannot really have been expected to have generously distributed meat amongst the whole clan with the perceptive intention of assuring normal brain development in their infants and children.

The 'gatherer' part of the human 'hunter-gatherer' heritage may have been more important for obtaining meat than the 'hunter' part. There was always carrion on the savannahs. Lions make a kill and satisfy themselves but then ignore a large portion of the available meat. Scavenging from carrion requires less skill than hunting live game but also yields less meat. The meat of grazing animals that are the prey of savannah carnivores has little body fat to improve the energy intake of scavenging hominids but the marrow of their long bones does contain fat. The big hunting cats rarely puncture the skull yet the brain itself could be an excellent source of nutrients and energy. Surely early hominids exploited this potential food resource along with the high nutrient quality of meat left on the carcass? Most likely they did, whenever they could. However, they had to withstand the risks of the meat being spoiled or of being attacked by the big hunting animals or other efficient scavengers such as hyenas. Assuming some people (presumably the young men) became highly skilled in locating carcasses and recovering the remaining meat, how much meat, bone marrow or brain would likely get transported back to the mothers, children and infants?

Thus, there isn't really a good case to be made to support the common premise that bigger brains or hunting were necessary for hominid survival as such. Indeed, nuts are richer in energy than most other foods including meat, so if more dietary energy were all that was needed to expand the brain, several primates as well as early hominids had a rich history of gathering fruit and nuts to rely on – no hunting was needed.

However, a nut gatherer has to constantly search for food on the ground or in the trees, which in itself is an energy-consuming process. The energy cost of living on nuts and fruit therefore compromises the goal of committing more dietary energy to brain evolution and fat stores. There are also several efficient quadrupedal primates that could easily compete for fruit and nuts with recently bipedal hominids.

Other dietary options besides fruit and nuts have been suggested. Digging for roots or raiding termite colonies adds variety to the diet. Still, the effort these activities require and the low energy yield of the termites or roots that are collected competes with the increased energy needs of expanding the brain. Therefore, it is doubtful that there would be much net energy gain by consuming roots or termites. Beyond these issues, the main problem with using fruits and nuts to explain increased fuel use by the brain is - how did only the hominid brain expand so markedly while consuming a diet so similar to that of other primates whose brains expanded to a much lesser extent, if at all?

Hunting fresh game or scavenging meat off carcasses would have been helpful but still insufficient to explain hominid brain evolution. Inherent in these concepts are the same problems – how, starting with the same primate brain as chimpanzees, did hominids alone develop weapons, which made them so much more effective at hunting? If hominids were eating a vegetarian diet broadly similar to that of other primates, how did they alone expand the brain to such an extent? Altered expression of some specific genes have been shown to markedly affect brain function in rodents (see Chapter 9) but there is still no suggestion that these genetic changes were involved in primate brain evolution let alone in more advanced cognition in hominids.

Why Survival Misses the Point

Starting with the Savannah Theory, most scenarios for human brain evolution implicitly or explicitly involve 'necessity': in order to survive when food resources were dwindling during the dry, early Pleistocene, hominids needed to make tools; in order to expand the brain, they required more dietary energy. In these scenarios, the connection between achieving brain expansion by solving the hunger problem is clear. Language and more sophisticated social dynamics were also supposedly needed to catalyze the hunting process and make it more efficient. This also required a larger brain. Larger brains required more energy, so a diet with a higher energy density had to be found.

Necessity puts the cart before the horse. Evolution occurs by natural selection, which operates purely by chance, catalyzed by opportune changes in both the genome and the environment. These changes are unplanned and sometimes undesirable. Sometimes the environmental change is too extreme and, like the dinosaurs, you don't survive. Or, like Darwin's finches, you are blown off course to isolated islands but you survive and propagate because the demands of your new environment, although different, are not too extreme. Sometimes the opportunity is too minor to induce evolution and, like alligators, for millions of years you don't change much, if at all.

Whatever it is, one cannot need to evolve a bigger brain (or anything else). Populations evolve but individuals survive or die. Pre-human hominids cannot have needed bigger brains, bipedalism, stone tools, or fat babies. Whether survival was or was not an issue, it could not cause human brain evolution. Concepts of human brain evolution based on enhancing survival miss the point.

Clearly humans did evolve bigger brains and have more advanced cognition than earlier hominids or other primates, so now we do need more dietary energy than chimpanzees if our brains are to develop and function to their potential. There is a crucial distinction between evolving big brains that subsequently need more energy, as opposed to first finding a better diet and then, as a consequence, evolving a larger brain, which then needs more energy to maintain its function.

Hominids cannot have needed to learn to hunt because they had to fuel their already bigger brains. Rather, they chose to hunt having already fortuitously evolved bigger brains that they were already nourishing and could continue to nourish without hunting. Brain expansion in certain hominids didn't occur under survival pressure. The conditions of primate life in woodlands or open savannah were not right for marked brain expansion or it would have happened to some notable degree in other primates.

Environmental Permissiveness

A dry, hot climate is widely felt to have been the key environmental pressure forcing habitat, lifestyle and dietary change in hominids that led to our present day human form. The pressure is seen as one for survival. For 2 to 3 million years, hominids are supposed to have become more adept at food procurement or would have gone extinct. In one form of another, harsh climate change is at the core of most theories of human evolution; this is the core feature of the Savannah Theory.

Two features about any crisis (survival)-based concept of human evolution are wholly unsatisfactory: First, by definition, crises don't leave much time, yet brain expansion in early hominids was fairly slow. Even though late Australopithecines were probably efficiently bipedal, it must have taken tens if not hundreds of thousands of years for *H. habilis* to become efficient toolmakers; how was *H. habilis* not only surviving but also evolving a bigger brain during the period before tool use became proficient?

Second, two of the three definitive attributes of humans (larger brains, fatter babies, bipedalism) seem to be the very antithesis of helping survival in harsh climatic conditions; large brains are not truly useful for survival for at least the first five years of life, and scarcity of food impairs brain function by reducing fatness in adults and babies alike.

Climate change was certainly happening in Africa 2 to 4 million years ago. Glaciation affected water levels, precipitation and temperature. Intermittent volcanic activity affected soil conditions, direction of

watercourses and the availability of many types of food. Not all
Australopithecines became *Homo*, so a harsher, drier climate might have
accelerated but was still not essential to converting Australopithecines
from a fruit- and nut-based diet to the meat-eating *H. habilis*. The
'robust' Australopithecines remained plant eaters yet survived for a
further 500,000 to one million years. If they didn't have to eat meat or
become large-brained to survive over this substantial period, survival
alone cannot be at the root of human brain evolution either.

Protein-energy malnutrition and specific nutrient deficiencies impair
brain development and function in all mammals in which this
relationship has been studied. If nutrient deficiencies affect rat and
human brain function in qualitatively similar ways, why is nutrition
especially important in human brain evolution? For the three reasons
outlined in Table 11.1. No other terrestrial species evolved this tidy
solution to the challenge of fuelling brain expansion. Nutrition of the
adult is part of the unique human solution to brain expansion because
sufficient excess maternal dietary energy is necessary in order to deposit
body fat on the fetus and neonate.

Table 11.1. Why nutrition was important in human brain evolution.

---

1. Presently, about one fifth of the world's human population consumes diets that
    inadequately support brain development.
2. The common nutrient deficiencies affecting humans can only be avoided by
    consuming shore-based diets.
3. Adequate body fat is a major determinant of normal brain development in infants.

Environmental pressure and selection pressure push the idea that
hominids had difficulty finding supper and were under pressure to make
'intelligent' decisions that almost instantly had a beneficial impact.
Environmental pressure means that, alone amongst the primates,
hominids had to almost instantaneously become proficient toolmakers
and had to be brave and wily enough to outwit lions and hyenas, so as to
protect the energy supply of their newly enlarging brains.

There is perhaps a better case to be made that environmental *permissiveness* rather than environmental pressure fits closer with the obligatorily blind nature of natural selection. What if some hominids stumbled into a habitat that provided better, more secure nourishment? What if dietary improvements that did not require tools or hunting permitted gradual brain expansion?

Shore-Based Human Evolution – A Vignette

Imagine that it is three million years ago and you're an Australopithecine teenager. That means you are little different from a chimpanzee and perhaps not quite as clever since chimpanzee brains have evolved too. You require all the things that other animals require – food, fresh water, shelter, and, eventually, a mate and reproductive success. However, the climate has been changing and it is getting dryer. The forest is shrinking and there is more competition for food. Like other primarily vegetarian animals of the forest, you are forced to look for new options, i.e. to become more competitive or go hungry. Some of your clan has perished from hunger or insufficient vigilance against attack by predators.

You are sitting on a log on a lakeshore or the edge of a large pond. You've seen fish before, some of which slow down and die seasonally after spawning. Raptors sometimes make off with one. You've seen bird nests containing eggs in reeds in the marsh. Perhaps you've even noticed a few clam or mussel shells in the mud. This time, though, you are hungrier than you can ever remember. Beyond the fringe of the forest, the lions have been more aggressively protecting their kills and hyenas are just too difficult to contend with. Chimpanzees are better climbers than you are and they have been getting to the ripest fruit just a little faster than you and your mates.

So you took a little more interest in those clamshells than before. There aren't many dangers and perhaps you are in a group so you are better protected and more willing to explore. You pick one up, wipe off the mud, bash it with a stick and realize that the juicy contents that spill out might just be edible. Indeed they taste good and you scrape around in the mud to find more. For a moment you forget about the effort of awkwardly scrambling up trees and grasping for fruit or nuts.

Circumstances like this would have been the humble beginning of the gradual hominid exploitation of one of the richest food resource on the planet – freshwater or saltwater shorelines. It started simply enough – as an option – stay in the woods and compete or look elsewhere. Chimpanzees stayed in the woods – you and your mates looked elsewhere. You didn't have far to look because you were used to drinking fresh water anyway. Initially, and for tens of thousands of years or more, you and your descendants weren't committed to any particular environment or diet.

Maybe more severe climatic change in some areas inexorably forced the use of shorelines as primary rather than occasional feeding grounds. Maybe in some locations, some Australopithecines started exploring shorelines without any sense of urgency at all, i.e. without being forced to do so by a drying or changing climate. Regardless, you and your clan gradually became aware that mussels, clams, crabs, fledgling or molting birds, crayfish, marsh plants, eggs, frogs, turtles, spawning fish and many other types of food were available in relative abundance on the shores of fresh water rivers, lakes and estuaries. They tasted pretty good and there wasn't much competition. You wouldn't have taken long to realize that most of these foods didn't poison you and didn't take all that much effort to catch or pry off rocks or out of the mud. Some were hard to catch but others were easier especially at certain times of the year. Some amongst your clan may have exploited the saltwater shorelines but would always have required ready access to fresh water.

You didn't know it at the time but you'd found the world's best source of brain selective nutrients and in relative abundance. Your taste for roots, nuts and meat would remain to be indulged as opportunity permitted. In the meantime, in a relatively short period of time you and your clan mates had started to explore and exploit a new but more reliable food supply.

How Could Shorelines Make Such a Difference?

Shorelines are richer in biomass than any other ecological zone on the planet. That means that food is commonly more abundant on shorelines than anywhere else. There also wasn't much competition – most other primates are low, mid-level or upper canopy foragers in the forest. On the shorelines, you could entertain yourselves trying to catch more elusive prey as you wished. Sometimes, these exploits would take you far afield and you got quite hungry and maybe some of your clan perished as a result. Mostly, though, you knew you could return to the same (or perhaps a different) shore-based camp.

The key point is that, for most of you, survival was not an issue. This is a crucial point in the evolution of your brain. You didn't depend on improved cognitive powers or better memory for survival. You'd found a place to eat and live, and finding your next meal was not a challenge. Your children could feed themselves. So could your aging parents and your offspring for thousands of generations to come. There wasn't any pressure to risk your life to be the great provider. The climate was dry but the massive lakes and rivers of east Africa never disappeared.

What did you do with this newfound freedom? You got a little fatter. So did your soul mate, and most of the rest of your clan, but especially your babies. You had time on your hands so you started to play with rocks, sticks and the like. Initially, these 'toys' held no particular importance to you. You and many subsequent generations were perfectly happy with this arrangement. Life was good. This was the start of a leisurely lifestyle that, three million years on, is now causing us heightened risk of health problems like diabetes, hypertension, atherosclerosis, cancer and obesity. Back then it was the fortuitous route to brain expansion.

Toys not Tools

One day, thousands of years after your seminal discovery, one of your descendants tried using the edge of a stone to cut something. Slowly they realized this worked better than using a fingernail or trying to tear it. The advantage of a sharp stone edge quickly became obvious because, unnoticed and unneeded by you, your brain was working a little more efficiently than in previous generations. The genetic potential for improved brain function had always been there in the primate genes you and previous generations inherited, but prior development of this potential was impeded by insufficient dietary availability of energy and brain selective nutrients. By stumbling upon an abundant and markedly improved diet, your generation would forever change that limitation on primate brain expansion. As a result of marginal and truly unnecessary improvements in brain function occurring over thousands of years, your descendants were the first to be able to imagine how such an object could useful as a cutter, scraper or digger, i.e. as a tool.

The first primitive tools were useful for play activities but they were not necessary for survival. Primates use sticks and stones to prod, bash or break things, but you were the first primate to start to realize the potential of tools to improve your life and the lives of your clan members. It hadn't mattered before and, indeed, it didn't matter now either. The insight that certain types of stone or wood might be useful crystallized gradually over many thousands of years, as the younger generations watched and learned from the older ones while little by little improving their ability to assess the collective experience with implements used as tools.

The key point is that, all the while, tools were unnecessary. They were optional. Effectively, they were playthings. Even recognizing their usefulness for a few simple routines of play depended on gradually improving memory, insight and cognition. These improvements were a consequence of increasing availability of dietary energy and nutrients to the embryonic and fetal brain, and more fat deposition both on the mother and on the fetus (and probably on the males as well).

The Venus figurines that appeared much later on show that artists noticed fatness in women. Cave images of animals were authentically proportional so it is reasonable to think that women proportioned as fat in figurines or etched into cave walls actually were fat. Teleologically, this would seem to have been necessary in order to have fat babies, so it is significant that there is a clear record of human fatness in the cave art of early humans.

Added dietary nutrients and infant body fat supplies were stimulating expression of genes affecting the timing and amount of brain cell development, resulting in increasing numbers of connections being made between brain cells. The 'hard drive' (brain size) was slowly getting a little bigger and the software (cell to cell connections through synapses) was becoming a little more sophisticated.

Instead of making a play tool, perhaps one of your distant descendants was pleased by the colour or pattern of something they were doodling with. So it was that art as self-entertainment began. Instead of cutting or colouring, perhaps you created a musical sound that amused everyone. You were playing and behaving just as children do today.

Young children don't think about tomorrow or even tonight; they live in the present. Their learning is opportunistic and they mimic. As any good teacher knows, the key to effective learning is a relatively stress-free environment lacking concerns about safety, nourishment or survival. A rich and reliable food supply afforded early hominids this same luxury.

The Infant is the Focus of Human Brain Evolution

The real focus of human brain evolution is the infant. An adult is less nutritionally vulnerable than is an infant, but to develop a healthy adult brain, one has to start out with a healthy infant brain. Intelligent children have at least the potential to become intelligent adults but mentally retarded children cannot become normally functioning adults. Humans can cope with a surprising range of physical infirmities but cannot easily overcome impaired brain development during infancy. Hence, it is essential to respect the need of the developing infant brain for a near constant supply of high quality food.

When food is temporarily lacking, the healthy term infant has abundant fat stores from which to draw for brain fuel, but low birth weight or premature infants do not have that 'energy insurance'. Neither do chimpanzee or other primate infants. Human infants and children are a steady drain on food resources long before they can contribute to the food requirements of the family, clan or village. Childhood is a time when one is most vulnerable and least able to fully exploit the innumerable advantages of a large brain. Most amazingly, how did hominid infants manage to be at the center of brain evolution while simultaneously investing an equal amount of metabolic effort in laying down an impressive layer of body fat as went into fueling their brains? How could hominids afford to concentrate a major part of the energy-demanding component of brain evolution in the fetus as it nears term and in the first five post-natal years?

The point is that hominids heading towards the human lineage were intent not on survival but on play. No one was under any pressure to find supper. The brain was starting to function slightly more efficiently and memory was improving. Imagination was born because of modification of the necessary genes, more reliable fuel supply, and because of a rich supply of brain selective nutrients was improving neuronal connectivity. More than anything, sufficient time was available.

Time is an essential element of successful play. As they get honed, skills learned during play can potentially be applied to survival strategies. By now your clan had evolved into what would much later be known as *H. habilis*, but your enhanced learning capacity was still associated with play. It was still very much opportunistic - useful, but not essential. The chance of making connections between specific objects and pleasant, rewarding functions was improving, but was never necessary. Crucially, there was no urgency to make these connections because you weren't hungry.

Eventually some of your descendants would have gone hunting. They would have hunted because they felt like it, or because the challenge intrigued them. Only rarely would they have gone because they needed to. By not needing to hunt, they would have had time to refine tools and skills in step with the minor, sporadic increments in intuition, memory and skill that were occurring almost randomly. It was in leisure and play not in hardship and strife that the human brain evolved. If supper was reliable and abundant, there was time to play and socialize. This is the blind, undirected process of natural selection at work. This led to a fundamentally different lifestyle, one of leisure and abundance, rather than grim-faced determination to overcome hunger.

Indeed, the leisure time enjoyed today by many aboriginal societies contrasts sharply with the supposedly 'nasty, brutish and short' lives of human ancestors. In some aboriginal groups, childhood in boys is allowed to extend to about twenty years old and is relatively free of the stress and obligation to search for food. These traditional groups measure affluence not in material possessions but in time available for story telling, music and dancing. Lumsden and Wilson note that the Ituri Pygmies and !Kung Bushman leave their children to occupy themselves by themselves much more than in supposedly more 'advanced' societies.

The children have no difficulty acquiring language, social and domestic skills by observation and mimicry.

Lumsden and Wilson note that the minds of these children *'flourish in these casual circumstances. They become sophisticated in speech, quotidian skills and tribal lore'*. Translation – they have time to learn because they have time to play, because finding tonight's or even tomorrow night's supper is not an issue.

Feckless, Irresponsible Pleasure

The shore-based environment allowed hominid adolescents and adults to be children for an extended period in life. Lewis Thomas, a prominent American pathologist, put the puzzle of human childhood succinctly. Invited to make a new list of the seven wonders of the modern world to replace the old, out-of-date original wonders, he named the human child as his Seventh Wonder because it seemed to him '... *unparsimonious to keep expending all that energy on such a long period of vulnerability and defenselessness, with nothing to show for it, in biological terms, beyond the feckless, irresponsible pleasure of childhood'*. That fecklessness, both for hominid children and adults, was a key ingredient in human brain evolution. With a reliable food supply, stimulation by thyroxine and nutritional tweaking of gene expression (see Chapter 14), play led to repeated activity that was optional to survival but which became reinforced because it was pleasurable and rewarding, and because certain hominids were developing sufficient brain connectivity to be able to evolve better memory.

Pleasure has a biochemical and genetic component, the rudiments of which exist in other primates like chimpanzees. The repeated activity that certain hominids found pleasurable and rewarding included acclimatizing to the water, chipping stone, serendipitously developing rudimentary musical instruments, appreciating the attractiveness of colour and design (art), and the sombre yet fulfilling reward of respecting the dead. These activities all occurred because the ability to experience pleasure was evolving and because of abundant leisure time. The fact that creative activities could and did emerge is incredible but the proficiency with which some, especially art, were developed, even

30,000 years ago, exceeds what all but a few humans, big brain and all, can do today.

Seeing the origin of these cognitive and creative developments in childhood and play may seem to trivialize the complex process of human brain evolution. That is not the intention. I am very serious about it precisely because play respects the interconnection of free time, memory, learning and the maturation of neuronal connections. Play has a vital role in normal childhood social and skill development. Any adult with an accomplished skill usually voluntarily develops an interest in that activity during childhood. Adults also know that it takes time – 10% inspiration and 90% perspiration, for talent to become skill. There is a tremendous amount of learning and repetition involved.

And that is just the effort that it takes in humans today, a species with an already enlarged brain that has unparalleled neuron-to-neuron connectivity. How could it have happened before the brain started to enlarge or while under duress to find supper? There isn't time to play or refine a skill if you are too occupied collecting food (or, in modern times, working excessively long hours). Conceiving of the origin of human cognition and creativity in play respects the fact that time is needed for creativity, innovative activity and skill development. With a lifestyle centralized on shore-based diets, time would have been available. The pressure of constantly searching for food was lifted, thereby providing the freedom and time to play.

The Shore-Based Scenario

The Savannah Theory has been taking a beating for many years. It was a framework but over time the weight of evidence didn't support it. The Savannah has become the Woodlands Theory largely because East Africa probably wasn't as dry as originally thought. Woodlands implictly involve more rainfall or groundwater to support the trees and are a wetter environment than the savannahs. Woodlands would require less hunting but would provide other tool-based food resources like tubers or roots underground.

Surely a non-specialist ape could evolve bipedalism and a big brain in this familiar environment? Not if you accept that the big brain and fat babies are the defining features of humans. Bipedalism in the woodlands - maybe. Primitive tool making - quite probably. Big brain and fat babies? Not likely, because the woodlands didn't offer enough geographic isolation or enough brain selective nutrients to evolve these features in any other large woodland primate, so how could they evolve in hominids? The genome has to be involved but it isn't different enough in humans and chimpanzees to account for these features, especially the big brain's vulnerability that is constantly exposed when shore-based nutrients, particularly iodine and iron are inadequately available.

The geological events of the past five million years in East and Northeast Africa are well known and clearly support the existence of large estuaries connecting the Red Sea, Gulf of Aden and perhaps other rivers from the Rift Valley reaching the Indian Ocean. This created large areas of various wetlands from South Africa to Saudi Arabia. Clearly volcanism and continental glaciation played havoc with the climate and landscape of East Africa on many occasions in the past two million years, altering lake drainage and river flow and, thus, habitability of preferred lake margins or river banks (or forests for that matter).

Bipedalism was an early feature of hominids so mobility to new, more hospitable sites became feasible. The ecology of marine or fresh water wetlands is compatible with survival of a crudely bipedal, non-specialist hominid 4 to 5 million years ago. Kathy Stewart has shown clear exploitation of wetland food resources, especially shallow water fish at several two million year old hominid fossil sites in the Rift Valley.

There is a two million year gap between the advent of bipedalism and the first clear evidence of organised fishing by *H. habilis* but earlier hominids could still have been exploiting other shore-based foods besides fish, including shellfish, eggs, turtles and plants. Shallow water fish and other shore-based foods provided a significantly better supply of brain selective nutrients than any exclusively terrestrial food supply. More than that, the fossil record shows that they were abundant in East and South Africa.

The *Shore-based Scenario* predicts that time was available, leading not only to fat deposition and fat babies but, from there, to uniquely human physical, mental and cultural skills. An abundant food supply makes it reasonable to hypothesize that shore-based life provided an excellent opportunity for leisure and play activity. The Shore-based Scenario isn't woodland and it isn't aquatic but it could well include opportunistic exploitation of woodlands or water as time, skill or ecological conditions permitted. What it does exclude is need (to develop tools to dig or hunt). The shores provided an excellent opportunity for the development of tools and hunting, but didn't require them; they allowed for the development of some aquatic or semi-aquatic features but didn't require them.

The Shore-based Scenario doesn't require the two phases of the aquatic theory (into the water and then out again). Neither does it involve the single dramatic phase of coming out of the trees onto the savannahs. It, too, is a single phase theory but involving ongoing dependence on the shores. Our ongoing dependence on shore-based nutrients for optimal brain development and to reduce the risk of chronic diseases (Morgan's 'scars of evolution') demonstrates that humans today are still in the shore-based phase of human evolution.

The Shore-based Scenario derives nutrients from shellfish, fish, shore plants, eggs, and small animals. By first approximation, this food selection has ample minerals and vitamins but, because of the non-plant sources, it appears low in fat, low in carbohydrate and high in protein. In fact, plants growing on the shores and in shallow water would probably make up the majority of the diet. They would be supplemented intake of shellfish, fish and other animals as opportunity or season permitted. Hence, carbohydrate would still be a major component of the diet, and fat would be fairly low but would be rich in omega-3 polyunsaturates, including pre-formed eicosapentaenoic acid and docosahexaenoic acid.

There is no *a priori* reason why the Shore-based Scenario wouldn't fit well within the dietary guidelines for human metabolism and survival. More importantly, there are advantages to the shore-based diet that no other diet selection could meet, because shellfish are the richest source of brain selective nutrients. Beyond diet *per se* is the overlooked but essential element of stationary, reliable, abundant food resources that would have reduced foraging time and increased leisure time, thereby setting the scene for play and learning.

The Savannah-cum-Woodland supporters consistently note that none of the supposedly unique 'aquatic' features were sufficient justification for an aquatic phase in human evolution. I agree that hominids didn't need to become fully aquatic but, to become humans, hominids did need a shore-based food supply. They still do. The best available evidence shows that we are still a shore-based species.

Chapter 12

Earlier Versions

The idea that certain hominids evolved through a shore-based phase on the way to becoming human is not new. From a chronlogical perspective, it all really got going in 1960, with Alister Hardy's article in *The New Scientist, Was Man More Aquatic in the Past?* Desmond Morris very briefly acknowledged Hardy's idea in his 1967 book, *The Naked Ape*, which is where Elaine Morgan first learned about it.

Morgan was Hardy's first unequivocal supporter, and she is still the most widely published. In a series of books starting in 1972 with the *Descent of Woman*, she popularized what became known as the *Aquatic Ape Theory*. Her support for and extension of Hardy's hypothesis continued with *The Aquatic Ape* (1982), *Scars of Evolution,* (1990), *The Descent of the Child* (1994), and the *Aquatic Ape Hypothesis* (1997).

Michael Crawford is a preeminent international expert on the role of docosahexaenoic acid in normal human infant development. While investigating the biochemical requirements for advanced brain function, he reported in 1972 that long chain omega 3 polyunsaturated fatty acids, especially docosahexaenoic acid, were important constituents of the mammalian brain. Before moving to London's Institute of Zoology, Crawford had spent several years in East Africa where the marked disparety in brain size relative to body weight between the large savannah herbivores and the primates had left a lasting impression on him.

I introduced Crawford to Hardy's hypothesis in 1981. Since Crawford was already aware that docosahexaenoic acid was uniquely prevalent in marine organisms and was an important component of the brain, he was immediately receptive to shore-based diets as a rationale for separating advanced human brain capacity from what was more typical of evolution on the savannahs, namely increased physical growth but relative *shrinkage* of the brain. With David Marsh, he published *The Driving Force* in 1989, which addressed the role of marine food in the evolution of *Homo aquaticus.* They explicitly supported both Hardy's original concept and Morgan's more detailed expansion of it.

My 1980 publication in *Medical Hypotheses,* entitled *The Aquatic Ape Theory Reconsidered,* started life in 1974 as a university term paper in a course taught by Leo Standing at Bishop's University (Lennoxville, Quebec). The goal of my paper was to report on the 'establishment' reaction to the publication of Hardy's first paper. When he published his *New Scientist* paper, Hardy was a well-respected professor and zoologist from Oxford University and Morgan had thoroughly and lucidly marshalled not just supporting evidence but the logical inconsistencies in the Savannah Theory.

By 1980, my take was that the Aquatic Ape Theory had generated little published reaction from the anthropology establishment and, save faint praise in Desmond Morris' *Man Watching*, none of what the experts said was complimentary. Hardy's hypothesis was not so much discredited by scientific argument as ignored; the few that deigned to comment on it all *knew* that hunting on the savannahs was the obvious path to human evolution. So, on the rare occasions it was mentioned, an aquatic phase was ridiculed as simply too complicated and unnecessary. The marked presence of body fatness in humans was a key element in Hardy's hypothesis but body fatness has long been a touchy subject for westerners so perhaps that doomed Hardy from the start.

With several colleagues including Michael Crawford, I published several papers since 1993 discussing the need for brain selective nutrients in human brain evolution. Our approach was to point out the significant metabolic and biochemical constraints on brain function as well as the evidence for vulnerability of the developing brain to nutritional deficit. We proposed human evolution was 'shore-based' because it met those metabolic constraints, did not require tool making or hunting for survival, and yet was not as committed to aquatic life as Hardy originally proposed.

Leigh Broadhurst, of the United States Department of Agriculture in Beltsville, Maryland, was a major contributor and lead author of several of these papers. She not only amassed detailed geological and environmental support for it but also obtained new data about the unique molecular confirmation of docosahexaenoic acid that helped explain its special role in the brain.

In an ongoing series of papers starting in 1985 in *Medical Hypotheses* entitled *The Aquatic Ape Theory: Evidence and a Possible Scenario*, Marc Verhaegen has systematically speculated about an aquatic role in the origins of several human attributes, notably dentition, speech, and brain function. Verhaegen launched his support for Hardy with detailed accounts of dental and skeletal features consistent with adaptation to an aquatic environment. Like Hardy, he proposed that streamlining of the human torso and boney protection of the ears and eyes represented a major commitment to aquatic life that included extensive and frequent diving. Verhaegen sees not only bipedalism but also brain enlargement and speech as arising in an aquatic environment. Unlike both Hardy and Morgan, who proposed that the aquatic phase occurred only in the period of the fossil gap (5 to 7 million years ago), Verhaegen has it running at least to the Neanderthal period.

What is New About the Shore-Based Scenario?

The Shore-based Scenario has evolved. By succinctly identifying several key physiological characteristics that seemed incompatible with the Savannah Theory, Hardy laid the initial groundwork. Morgan and Verhaegen endorsed and expanded the Aquatic Theory around further dissatisfaction with explanations for unusual human physiology and anatomy. Morgan added seminal concepts about both the importance of the child in human evolution and the physiological penalties or *'scars of evolution'* that all species pay, no matter by which path they evolve.

The disregard by paleontologists and physical anthropologists for the legitimacy of Hardy's fundamental observations on the uniquely fat and stream-lined, bipedal human form is what lit the flame for me. It rankled that no evolutionist would say that Hardy's observations were valid even if his interpretation was radical. Crawford and others showed that the link between human brain function and docosahexaenoic acid was vulnerable to dietary inadequacy of omega 3 polyunsaturates.

Elaine Morgan intrigued me with *Descent of the Child,* and Michael Crawford amassed compelling evidence for the importance of docosahexaenoic acid in brain development. Crawford had also written about how widespread inland iodine deficiency must have been an impediment to savannah-based human evolution. These 'earlier versions' of the Shore-based Scenario each emphasized different aspects of human physiology and metabolism, and proposed different degrees of hominid commitment to semi-aquatic life.

On the basis of this mosaic of observation, evidence, and controversy, I add that there is more to the shore-based environment than a nebulous 'better quality diet' needed to fuel brain expansion. Throughout the 1980s, my research at the University of Toronto seeded the idea for a cluster of brain selective nutrients. We then discovered several new attributes of baby fat that seem to contribute to insuring brain development : First, we showed that body fat in a baby born at term contains sufficient docosahexenoic acid to meet the brain's needs for several weeks without depending on docosahexaenoic acid supplied by maternal milk or by synthesis within the baby itself. Second, most but not all of the fatty acids in baby fat are fuel insurance for the brain. Some of the fatty acids will be made into ketone bodies which become the main building blocks of lipids, e.g. cholesterol, that are arguably even more important in the brain than docosahexaenoic acid.

In essence, the insurance provided by fatty acids in human baby fat releases three important constraints that limit brain development in other mammals, including primates: they are an important fuel store; they provide a ready-to-use supply of docosahexaenoic acid; and, they provide an abundant reserve of ketone bodies for synthesis of brain lipids. However, the effectiveness of this insurance absolutely depends on the amount of body fat at birth. With prematurity or low birth weight, the insurance reserves are markedly reduced and the consequent vulnerability of the brain's development is immediately revealed. This work formed the basis for my '*Survival of the Fattest*' article published in 2003 in *Comparative Biochemistry and Physiology*, in which fat babies were proposed as a key prerequisite to evolving big brains.

My other original contribution to this discussion is to reject the 'necessity' of a better quality diet as the primary force in human brain evolution. Once the brain was larger and more advanced, a better quality diet was necessary. However, the fundamental constraint of natural selection is that nothing evolves (or can be retained) because of necessity. Evolution of the human brain became possible by a change in lifestyle owing to fortuitous exploitation of the shores. To be sure, the changed lifestyle on the shores markedly improved nutrition but food abundance also left pre-human hominids with plenty of time away from the monotonous and sometimes arduous demands of food gathering. With free time on their hands, play emerged which was the forerunner of tool making and creativity. More free time expanded socializing, which was the forerunner of language and culture. Free time also slowed down activity, which was the forerunner of fatness.

Ongoing vulnerability is the outstanding consequence of commiting the brain to expansion, greater complexity and dependence on increased intake of energy and brain selective nutients. The World Health Organisation and other agencies have reported that vulnerability directly compromises brain development in about one quarter of the world's population. Until eighty years ago, iodine deficiency caused impaired brain function ranging from mild retardation to full blown cretinism in substantial areas of the USA and western Europe. It may also have contributed to gradually declining human brain size over the past 30,000 years.

Jerry Dobson of the Oak Ridge National Laboratory in Tennessee suggests iodine deficiency may have been an important factor in the limited creativity of Neanderthals and possibly in their eventual demise (see Chapter 13). Premature and low birthweight infants face formidable challenges to attaining normal neurosensory development, in large measure because of issues related to nutrient and energy supply. Serious concerns are also being raised about whether the higher prevalence of depression and psychological, psychiatric, and neurodegenerative diseases in many 'westernized' countries is related to lower intake of several brain selective nutrients, notably omega-3 polyunsaturates: humans have not escaped this ongoing vulnerability.

This is the context from which my version of the Shore-based Scenario emerged. The ground-breaking contributions of Hardy, Morgan, Crawford and Verhaegen clearly fertilized my ideas but the predominant roles of baby fatness, brain selective nutrients and leisure time are where my contribution is original. Let's backtrack and lay out in a little more detail some of the features of those initial formative contributions.

Sir Alister Hardy, FRS

When I first mentioned Hardy, he was simply Alister Hardy. Now I've called him Sir Alister Hardy, FRS (Fellow of the Royal Society). The added credentials came after he had formulated his novel ideas about an aquatic phase in human evolution but before he reported them publically or published them. This delay was no coincidence as he readily acknowledged. Perhaps knighthood was not on his mind but an FRS, the top academic recognition in the UK, certainly was.

Hardy recognized the irony that if he publicly presented his radical hypothesis before receiving his FRS, it would probably be met with skepticism from influential people and might well derail an otherwise healthy shot at the coveted FRS. He therefore waited, just over thirty years, actually. With the FRS finally in hand, he chose a low key public lecture to announce his idea. Distortions of what he said in the subsequent media reports made it essential to rapidly publish his own version of what he had said in the lecture. The result was his now famous article in *The New Scientist – Was Man More Aquatic in the Past?*

What Hardy said in that article was that representatives of most groups of animals, extinct and living, have had aquatic phases in their evolution. For some, the transition to aquatic life has been permanent but others continue to occupy both land and water. Several features of humans seemed to Hardy to be empirically easier to understand if pre-human evolution had also involved a '*semi-aquatic*' phase in which searching for food while wading gradually gave way to locomotion by swimming. Gradual improvements in breath-holding would allow shellfish and crustaceans to be caught by diving further from shore.

The alignment of hair tracts is different in humans than other primates. The reduced amount of hair and their alignment with the flow of water seemed to Hardy to facilitate swimming. Bipedalism seemed more plausible if one were toddling and then wading into deeper water because boyancy helps support a vertical body position. Body fat is distinctively absent in primates but present in aquatic mammals. Unlike other primates, the thumb opposes the fingers on the human hand and this makes it easy to pick up or hold things. The fingers are sensitive enough to grasp and identify items even without being able to see them.

If hominids were exploring the water's edge, the selection of crustaceans, molluscs, turtles, eggs and edible plants on the shores would naturally have been noticed, tested and exploited. Hardy recognised that if sea levels had risen over the last several million years, such a semi-aquatic phase in human evolution might explain at least part of the lengthy gap in the hominid fossil record prior to the earliest Australopithecines. He concluded that his *'thesis is, of course, only a speculation – an hypothesis to be discussed and tested against further lines of evidence'*.

The New Scientist is at the interface between the peer-reviewed reports in the primary research literature and the popularization of science in more fluid, digestable, but opinionated form. I suspect that Hardy chose *The New Scientist* because, alarmed at some of the extreme interpretations of his lecture that found their way in the popular press, he wanted a rapid, respected but also widely disseminated printed medium to clarify his ideas before they became permanently misconstrued.

He claimed that his original intention was to publish a book providing a more complete version of his hypothesis but this never happened. If he had supported his ideas in more detail, this might or might not have enhanced his stature amongst physical anthropologists. However, publishing his first and only widely distributed article on this subject in a format unreviewed by his peers certainly did not improve the odds of him being taken seriously. That it was a radical concept to boot more or less guaranteed he would be ignored or criticized by anthropologists.

Seventeen years after the first, Hardy published a second article, *Was There a Homo Aquaticus?*, in *Zenith*, the magazine of the Oxford University Scientific Society. He added little of substance to the original idea but noted in his paper that another well-known zoologist from Oxford, Desmond Morris, and a TV script writer from Wales, Elaine

Morgan, had by then written about his hypothesis. Morris' 1967 book, *The Naked Ape,* noted simply that an aquatic phase *'would not have seriously clashed with the general picture of the hunting ape's evolution out of the ground ape'.*

Notable in Hardy's second article was his affirmation from his original article of an extremely long fossil gap, during which he proposed hominids had a semi-aquatic life. A quotation from that article sets the context of his idea : *'I must make it clear that I do not suppose man spent more than perhaps five or six hours in the water at a time.* Homo aquaticus *left the sea (or lake) a very different creature from when he first entered it. Now with a hairless body, subcutaneous fat ..., a knowledge of making and using tools and, above all, the erect posture, he might well be called a new species of man : indeed the ancestor of* Homo erectus.

This was a shore-based proposal that seemed to give a plausible account of the evolution in humans of bipedalism, body fat and and tool making during a period when hominid fossils had, at the time, been hard to find, thereby leaving these issues unanswered. It is clear from this quotation that Hardy saw the aquatic phase as transitional and that, once fully bipedal, hominids in the form of *H. erectus* were firmly back on dry land. Whether he imagined that *H. erectus* and early humans would have remained shore-based is unclear.

Elaine Morgan

Hardy was the creator of the original aquatic hypothesis but Elaine Morgan became and remains its heart and soul. Actually, she has pointed out that eighteen years before Hardy, a German professor, Max Westhofer, wrote a book chapter suggesting much the same hypothesis. However, his version was poorly circulated outside Germany and Hardy had been unaware of it.

Morgan added form and substance to the aquatic hypothesis. By the time she wrote *The Aquatic Ape Hypothesis*, popular dissemination of the original idea had transformed it into the *Aquatic Ape Theory*, a term that, with typical modesty, she resisted because too much of it was still controversial. Morgan discussed the good, the bad and the ugly of human evolution. *The Descent of Woman* was her first assessment of the problems with the savannah theory. In it, Morgan suggested that human evolution, whatever its path, was as dependent on female as on male evolution. She dismissed the widespread notion of human evolution based on pair bonding in which males provided food and were rewarded with sex.

In *The Aquatic Ape Hypothesis* and in *Scars of Evolution,* more than anyone else before or since, Morgan discussed a wide range of anatomy and physiology peculiar to humans. She defended the aquatic hypothesis but still accepted its limits – many human features seem more plausible, but cannot be exclusively explained, on the basis of shore-based evolution.

Above all, she emphasized that evolution affects the whole organism : bipedalism may have certain advantages ensuring its adoption but that doesn't mean there aren't some disadvantages too. Once one is bipedal and the viscera are stacked vertically with the heart lower than some organs but higher than others, back trouble, and more difficult birthing, i.e. Morgan's 'scars of evolution', are easier to understand. In his introductory book, *Evolution,* Morton Jenkins summarized it neatly: *'something must have favoured the erect posture ... but, whatever it was, we pay for it today with hernias, pot-bellies, difficult childbirth and back problems'.*

One of the most outstanding potential disadvantages to being human is being a human baby. The degree and duration of extreme vulnerability are certainly more marked than in other primates. If the infant is inadequately nourished, the risk of impaired brain development is lifelong. Curiously, although infant formulas represent the modern 'improvement' in infant feeding, Alan Lucas of the Institute of Child Health in London points out that until recently, many such formulas were nutritionally inadequate and had a real risk of limiting IQ by as much as 10 points in the first ten years of life.

In *The Descent of the Child*, Elaine Morgan focussed on the unique evolutionary situation of the human child. As Morgan sees it, and I agree, the disadvantages of being a human child were made tolerable by shore-based evolution. The combination of fatness, large brain, rapid learning of speech, and developmental vulnerability do not coexist elsewhere in the animal kingdom.

The present book is really an extension to *The Descent of the Child*. It lays out in more detail how the developing brain is vulnerable to nutrition and why body fat is the essential currency of the extraordinary brain development in human infants. Whatever the advantages body fat may have for bouyancy and insulation in infants, increased energy availability to the mother and child are the *sine qua non*. Body fat in neonates and infants starts to be laid down early in the final third of pregnancy and becomes the most expensive part of fetal growth towards normal term. Maternal energy restriction virtually guarantees less body fat and thinness at birth, especially if birth is premature.

A lifestyle affording a reliable energy intake that includes an abundant source of brain selective nutrients is necessary to avoid developmental delay in humans. Woodland or savannah settings do not meet these criteria but the shorelines do. Bouyancy and insulation weren't the reason for neonatal fatness but became a fortuitous result of dietary richness available from shore-based existence that then led to greater tolerance and exploration of water throughout the human lifespan.

Michael Crawford and the Land-Water Interface

Michael Crawford's crucial insight is that humans did not actually evolve larger brains. He doesn't question that humans have larger brains than other primates, but maintains that land mammals including other primates *lost* relative brain size as their bodies became bigger. Small mammals have relatively large brain size, several on the order of 2% or more of body weight, which is similar to humans.

Early hominids like *A. afarensis* had brains occupying about 1.7% of body weight before the split between the 'robust' and 'gracile' Australopithecines. The robust *Paranthropus* line lost relative brain size as did the eventual modern non-human primates. However, the *Homo* line came close to maintaining the same relatively high brain to body ratio of 1.7-1.9% right through *H. Neanderthalensis*. With the emergence of Cro-Magnon and humans, only a small proportion of the final gain in brain to body weight ratio is attributable to an absolute gain in brain size; most was due to decreased body size in comparison to Neanderthals.

Other primates such as the gorilla, but especially the savannah herbivores like the elephant and rhinoceros, lost relative brain size because they became physically immense. Their brains were unable to keep pace with their physical growth because the savannah offers not only a paucity of brain selective nutrients but also food selection with low energy density. Amongst the herbivores such as the pig and sheep, those that are wild have bigger brains than have the domesticated varieties, even though the domesticated ones generally have access to more dietary energy because they are being fattened up for market.

More than any one else, Crawford was responsible for identifying the limiting nature of the savannah food chain for the brain, especially the absence of arachidonic acid and docosahexaenoic acid, in all diets except those of the carnivores. He points out that, among large mammals, only the dolphin has keep close to the human pace in brain size, brain proportion and brain complexity, and that the dolphin is in a marine environment. My own ideas concerning the role of long chain polyunsaturates for brain function were formed and have been maintained during our close working association over the past twenty years.

This puts Crawford firmly in the Hardy-Morgan camp at the *land-water interface* in explaining how humans evolved larger, more complex brains. He firmly believes we remain today as vulnerable as ever to an inadequacy of shore-based nutrients. Hence, our dependence on this unique ecological zone is far from over.

Marc Verhaegen

Over the past twenty years, Marc Verhaegen has meticulously gathered and published evidence supporting the Aquatic Ape Theory. In several detailed papers, he attributes numerous human or pre-human skeletal features to semi-aquatic life spanning *H. erectus* to *H. Neanderthalensis*. These features include a thicker, elongated skull, heavier skeletons especially the long bones of the legs, spinal and pelvic adaptations, enlarged sinuses in the maxilla (behind the nose), and external boney protection outside the ears and above the eyes.

In reference to the large human brain, Verhaegen has developed a detailed case suggesting human cranial anatomy and localized brain enlargement is a consequence of previous adaptation to semi-aquatic life involving extensive wading and diving. Areas of the brain originally controlling breathing and mouth and neck muscles would have, by exaptation, been recruited to become Broca's area, which is used for motor control of speech in the mouth and throat. Like Broca's area in the precentral gyrus, Wernicke's area or the primary speech centre in the post central gyrus, is unique to humans. In fact, Verhaegen notes that they are connected by a uniquely human nerve tract, the *arcuate fasciculus*. Together these two brain areas contribute significantly to expanded brain size in humans.

Although barely identifiable from indentations in the skull of at least one *H. erectus* speciman, these features did not occur in the smaller and simpler Australopithecine brain, nor do they exist now in the chimpanzee, or in brains of other primates. Therefore, Verhaegen suggests that wading might have begun earlier but swimming and diving only became fully developed with the emergence of *H. erectus*. Verhaegen would place *H. erectus* in a more salt water coastal location whereas Neanderthals would have swum and dived in freshwater. Hence, contrary to the views of Morgan and Hardy, he feels the *Homo* lineage and not the Australopithecines were the fully developed *'aquatic apes'*.

Like Morgan, Verhaegen recognised that the return of *Homo* to a predominantly terrestrial existence would carry with it 'scars of evolution' particular to humans. His 1987 paper in *Medical Hypotheses* details several such scars, which fall into categories related to disorders of respiration (hyperventilation, spasm of the larynx and bronchi), diseases of the skin (seborrhea, dandruff and acne), and problems of upright posture (hernia, varicose veins and osteoarthritic joint disease). In fact, this paper stimulated Morgan's 1990 book, *Scars of Evolution*, which covered different aspects of the same subject.

When did the Aquatic Phase Occur?

In both his brief accounts, Hardy was specific about when he thought hominids were semi-aquatic : for him it was in the period of the 'fossil gap' that ended about five million years ago. In both papers, the aquatic phase occupied the whole fossil gap but he did not extensively defend his reasoning beyond noting that it would account both for the emergence of bipedalism and the absence of hominid fossils during that period. In *The Aquatic Ape Hypothesis,* Morgan supported Hardy's time line but shortened it : '*the hypothetical aquatic phase of the ancestral apes during the fossil gap would have been brief, a matter of two or three million years*'. Clearly, for both Hardy and Morgan, the aquatic phase was the essential prelude to bipedalism which was the first genuinely hominid feature, predating the main phase of brain enlargement by 2 to 3 million years.

In discussing the principal elements accounting for the evolution of bipedalism in hominids, Verhaegen explicitly supports the Hardy-Morgan hypothesis. However, he also describes a semi-aquatic stimulus for brain expansion in *H. erectus*, especially as it related to the origins of human speech. Accelerated brain expansion did not start until the appearance of the fully bipedal *H. rudolfensis* and *H. habilis*, both about two million years ago.

Verhaegen also attributes several features of Neanderthals to semi-aquatic life. Since he agrees with the semi-aquatic phase driving bipedalism, Verhaegen clearly extends the semi-aquatic phase well beyond the 4 to 5 million year old limit set by Hardy and Morgan to a period only tens of thousands of years ago. If, as Hardy and Morgan claim, the semi-aquatic period ended 4 to 5 million years ago, it was over before substantial brain expansion had even begun. It could only have contributed materially to brain expansion if some hominids had remained shore-based.

A semi-aquatic phase beginning at least five million years ago may well have been able to contribute to the slow, early stages of hominid brain expansion. Still the question remains - if hominids were semi-aquatic and consuming a shore-based diet during a two million year period that ended five million years ago, why was the start of the main period of brain expansion delayed for at least two million years, i.e. until *H. erectus* evolved?

Suppose for argument's sake that the situation is simple - that there are two distinct theories of human evolution – the Savannah Theory and the Aquatic Theory. Suppose also that both have their hardliners and their softliners. Thus, some savannah supporters are more literal about the savannahs and see hominids initially scavenging antelope carcasses but then progressing fairly directly to hunting large animals, while others see a more wooded environment with varied subsistence including fruit, nuts, termites, etc.

Starting with Hardy himself, some aquatic theorists are also more hardline and describe streamlining and hair tract patterning of the human form as arising due to a significant commitment to swimming and diving. Others propose a more moderate view based on wading in marshes, with greater emphasis on food gathering from the shores and shallow water rather than from deeper areas requiring diving. Some aquatic supporters focus on a fresh water milieu; others feel sea marshes and shorelines held less competition or risk from predators and so were more compatible with human evolution.

Hardliner or softliner, everyone with an interest in human origins has an opinion about the plausibility of one or other of these two theories, even if they only focus on the opposing theory's most glaring weaknesses without recognising their own. Let's therefore be clear about a couple of things : First, the situation is not simple; there is a single version of neither the Savannah nor the Aquatic Theory. Rather, depending on the proponent, descriptions including their time frames are both fairly fluid. Second, in both theories, 'need' to adapt to a new environment for survival is still implied as a driving force. Key questions about human uniqueness expose the weaknesses or the strengths of both theories : How did the hominid brain expand without the support of shore-based nutrients? How did hominid babies get fat without a sedentary lifestyle? How did bipedalism arise in hominids and did it occur quickly or slowly? Why would tool use have arisen? Why are many hominid fossils apprently distant from shorelines?

Furthermore, there are contentious issues about lineage, age, physical form and habitat during human evolution that interest many anthropologists more than settling whether human origins were strictly aquatic or savannah. How the brain got larger is less important to some people than knowing which areas of the brain enlarged and in what order, or what cultural skills or artifacts accompanied brain enlargement.

For these individuals, one or other theory is a good enough working framework for their particular specialization and getting to the bottom of either is not their main subject of interest.

Therefore, focussing on generic versions of the savannah versus aquatic theories polarizes the situation unnecessarily. In fact, many paleoanthropologists now reject the original version of the Savannah Theory just as they always rejected the Aquatic Theory. Currently, a softer, woodlands version of the Savannah Theory holds the most sway but the aquatic supporters still tend to dismiss anything non-aquatic as the Savannah Theory.

Compared to the Hardy-Morgan version of the Aquatic Theory, the Savannah-woodland Theory starts out from a position of strength because basically everyone studying human evolution learns it. With few exceptions, most leading contemporary physical anthropologists are savannah-woodland supporters. One of those exceptions is Professor Philip Tobias of Witwatersrand University in South Africa, who has on several occasions publicly said that it is time to lay the Savannah Theory to rest. Nevertheless, relatively few students of anthropology ever hear about the Aquatic Theory and when they do it is usually only in a cursory or perjorative context. This is in part because aquatic ape supporters are mostly non-specialists in human evolution who have little clout in anthropology circles.

***

Hardy's point was that humans have several attributes quite unlike those of other primates but in common with aquatic mammals. He asked whether human ancestors could have been more aquatic in the past. The logic of aquatic features, therefore an aquatic past, was arrestingly simple, but intellectually also too radical. Hardy's hypothesis is a legitimate framework for scientific enquiry. The evidence for what might be called 'predominantly aquatic attributes' in humans is impressive. Occurrence of these attributes in a species said to have evolved exclusively on the land is, or should be, unsettling.

Hardy's insight about aquatic features was a flash, a serendipitous connection. It was something he ignored for long enough, but he remained captivated by its simplicity. In retrospect, what is perhaps surprising is that although Hardy held off publishing for so long he amassed so little evidence during that interval. Even after publication of his short controversial papers, he did not subsequently publish detailed support for his hypothesis. It is therefore understandable that physical anthropologists, for whom hard, conclusive evidence is so difficult to come by, would dismiss an insightful concept because it was given so little substance by its originator.

In addition, animals with no known or even plausible aquatic past are at home in the water, e.g. several strains of dog. A diving reflex can also be experimentally induced in dogs. Hence, physiological or anatomical features most easily interpreted as representing adaptation to water, including possession of a diving reflex, are circumstantial and are not in themselves sufficient evidence for aquatic evolution. It is not hard to see how the baby has been thrown out with the bath water.

It is therefore time to turn to the harder geological and fossil evidence for shore-based human evolution.

Chapter 13

The Evidence

One would one expect to find the fossils of shore-based hominids on the shores. The problem is that the position of the shorelines changes when the water level changes. If the water level goes up, fossils on the shores go under water and become inaccessible. If it goes down, they end up on exposed sites or washed downstream by rivers, and much of the context of where and when the individual died is lost or gets confusing. This is a major problem with the Aquatic Theory – many anthropologists have dismissed it primarily because they claim there is no supporting fossil evidence in geographic regions where hominid evolution was occurring.

Despite the general impression one gets from the literature, quite a few reports by some highly respected physical anthropologists place early hominid remains near fresh water lakes in East Africa. In referring to hominid fossil discoveries on the shores of Lake Turkana in northern Kenya, the title of Leakey and Lewin's book - *People of the Lake* – seems to make the connection obvious. In *Becoming Human*, Ian Tattersall refers to the *'riverine forests'* at Hadar and the *'forests surrounding an ancient lake'* at one of the most famous hominid fossil regions of all - the Olduvai Basin in northern Tanzania. In *Java Man*, Carl Swisher of the Berkeley Geochronology Group refers to 1.5 million year old *H. erectus* encampments at a *'soft sandy bank on the bend of a small stream'*.

In *Lucy – The Beginnings of Humankind,* Donald Johanson and Maitland Edey describe the Hadar area in northern Ethiopia 3 to 4 million years ago as having been a lake bordered by plateaus. At the time, this area underwent considerable volcanism, earthquakes and

243

climate change. Rivers coming off the plateau carried sediments that gradually filled in the lake and, in the process, covered the fossils. Basalt layers from nearby volcanoes and faulting both changed the topography and direction of river flow and sediment deposits on several occasions. A rainy climate 3 to 4 million years ago was followed by extensive drying which has resulted in the virtual desert that exists there presently.

Johanson and Edey clearly indicate that *'hominids lived along the lakeshore throughout this entire period'* (4 to 2.6 million years ago). Elsewhere in *Lucy*, they describe the areas around Omo and Turkana in Kenya and Afar in Ethiopia as *'lush lake regions swarming with game, laced with winding rivers and stands of thick tropical forest; now near desert. Laetoli was drier than they were then and greener now... there are several lakes in its vicinity'*. Later on, they mention the *'lake shores, rivers and forests of the Afar landscape three million years ago'*.

Curiously, but perhaps most importantly, these and other descriptions don't mention possible exploitation by hominids of these lush shoreline settings for food. Richard Klein of Stanford University noted that hominids in south Africa that pre-dated *H. erectus* *'camped on stream banks or on channel bars, on lake margins, near springs...'*, but he was skeptical that they were exploiting these shores for food. Did they camp at these lake and river shores simply for fresh water while seeking food elsewhere, or did they camp there because they were eating shore-based foods?

The Geological Evidence

Many of the earliest Australopithecine fossils have been found in present day northern Kenya and Ethiopia. In fact, most fossils that predate the first global hominid wanderer, *H. erectus*, have been found in a string of sites which all have one thing in common – the five thousand kilometer long *Rift Valley* of East Africa. The southern end of the Rift Valley is near the Zambesi River in Mozambique (Figure 13.1). From there, it runs north, splitting around Lake Victoria. The western split predominates into northern Uganda and large sections of it are now occupied by big lakes including Lakes Tanganyika and Malawi.

The eastern split is indistinct through Tanzania but predominates again as the Rift Valley continues to cut north through Kenya into southern Ethopia where it gradually widens out to the northeast. Once as far north as present day Addis Ababa in central Ethiopia, it continues to spread out with the western edge heading north to the Red Sea and the eastern edge heading east through Somalia to the Gulf of Aden. This widening of the northern part of the Rift Valley is known as the *Afar Triangle*.

Continental drift and seafloor spreading characterize the geological origins of the Afar Triangle. About thirty million years ago near the end of the Oligocene and beginning of the Miocene epoch, the continent of Africa, or more correctly the African plate was slowly pushing the Arabian plate into the Eurasian plate. Over the next fifteen million years the dynamics of this collision involved upwards and downward heaving of large blocks of continental crust, which gradually shaped the Afar region, Red Sea, Gulf of Aden, and the huge lakes of East Africa. In short, these events created the essential geological elements of the Rift Valley that exist today.

During this period, the Mediterranean and Red Seas were at times connected and at other times separated. Both dried out several times. Towards the end of the Miocene, perhaps 7 to 8 million years ago, the Red Sea was separated from the Gulf of Aden by the *Afar Isthmus* and was a long, narrow gulf open only to the Mediterranean Sea. The Afar Isthmus was an important land bridge between Africa and Eurasia by which several animal migrations occurred during the late Miocene.

By the beginning of the Pliocene about five million years ago, the Red Sea was no longer a gulf of the Mediterranean but was instead open at the other end to both the Gulf of Aden and a significant stretch of the northern end of the Rift Valley as well. Continuous water between the Red Sea, Gulf of Aden and Rift Valley cut off present day central Somalia from the Ethiopian escarpment to the west and the Arabian peninsula to the north.

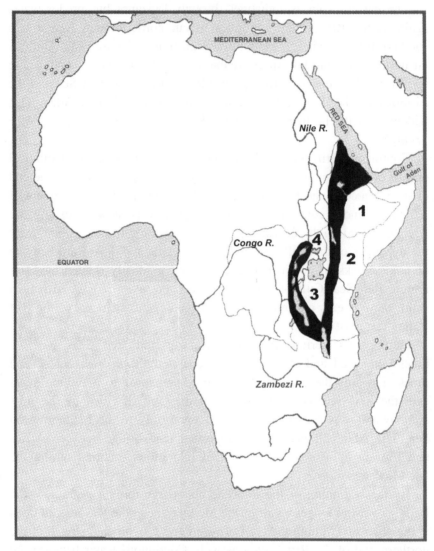

Figure 13.1. The Rift Valley of East Africa (shown in black). The main countries through which it passes include Ethiopia (1), Kenya (2), Tanzania (3) and Uganda (4).

In the Miocene, the *Danakil Horst* was a major feature of the uplifted area that created the Afar Isthmus at the present day junction of the Red Sea, Gulf of Aden and Afar Gulf. At that time, the Danakil Horst formed the northern boundary of the Afar Triangle and the southern boundary of the Red Sea. This area is now known as the *Danakil Alps* which run about 500 kilometers northwest to southeast along the southern edge of the Red Sea.

Along the lower elevations where its northwestern side faced the Ethiopian Escarpment, the Danakil Horst was filled with salt deposits that were part of the Red Sea. This area is called the *Danakil Depression*, which was located in the middle of ongoing tectonic activity. Sedimentation throughout the Miocene created deposits known as the *Red Series*, which are found on both sides of the depression. At the time these deposits were forming, the Danakil Depression was a lake or bay extending from the Red Sea. Periodic volcanic activity spread layers of lava over the lake bottom's sediments. The periodic drying out of the Red and Mediterranean Seas probably occurred in the Danakil Depression as well.

Five to six million years ago, the Danakil Horst twisted and tilted resulting in the waters of the Red Sea and Gulf of Aden joining and occupying a major part of the Rift Valley. Thus, according to Leon Lalumiere, previously of the Naval Research Laboratory in Washington, who first developed this scenario, the Danakil Horst became *Danakil Island*, which was about 500 kilometers long and about 75 kilometers wide. Danakil Island may still have been connected to present day Somalia but, because of massive basalt layers from volcanic activity, the neck of land connecting them was devoid of vegetation so the Danakil Horst was at least a 'biological island' on which Lalumiere proposes some pre-Australopithecine apes could have been marooned.

The region around Danakil Island has yielded hominid fossils. Still, very early hominids have been found at other locations further south in Kenya and further west in Chad. Hence, locations with the the geology and ecological isolation of Danakil Island may be generically important to hominid speciation and could have been duplicated elsewhere in eastern and northern Africa.

Leigh Broadhurst, Michael Crawford and several other collaborators including myself developed a similar scenario but instead of focussing on the single, fairly restricted Danakil Island site in Lalumiere's scheme, suggested that the shores of massive lakes of the Rift Valley were the sites at which hominid bipedalism evolved. Broadhurst notes that the junction of the present day Red Sea, Gulf of Aden and northern Rift Valley are the only present day example of failed or *'proto-oceans'*. During the late Miocene and early Pliocene, these bodies of water occupied at least a million square kilometers. They formed in sunken basins that occurred during the faulting that formed much of the Rift Valley 8 to 9 million years ago.

Today, the remnants of these proto-oceans include Lakes Victoria, Turkana, Tanganyika and Malawi, several of which have deep water geological features peculiar to oceans. At about 70,000 square kilometers, Lake Victoria remains the world's largest tropical lake. It dried out during the late Pleistocene and reformed about twelve thousand years ago. Like the others, its level still varies markedly with temperature and rainfall.

Broadhurst distinguishes between the different climates and outcomes of hominid evolution on the eastern compared to the western sides of the Rift Valley. She proposes that the western side remained more humid and forested, conditions which continued to sustain the great apes. In contrast, the eastern side became drier and less forested due to ongoing volcanic activity and uplift. Despite the dryness of the Pleistocene period, there were still abundant lakes, marshes and rivers, including Paleo Lake Hadar in the north, and Paleo Lakes Turkana and Olduvai further south in Kenya. Contemporaneous Australopithecine fossils have also been found near lakes much further west in present day Chad. These lacustrine and riverine conditions supported the evolution of hominids and, later, their transition into humans.

The Ecological Evidence

Geological evidence in the Rift Valley clearly shows that the sediments where numerous Australopithecine fossils have been found were adjacent to or formed by lakes, marshes or rivers. Derek Ellis is a marine ecologist from the University of Victoria in British Columbia who has addressed whether the ecology of such habitats would have been able to support the evolution of bipedalism and the speciation of hominids. As with others seeking a more viable framework for the antecedants of human evolution, his premise for investigating the plausibility of hominids evolving bipedalism in a shore-based ecosystem arose from dissatisfaction with the Savannah Theory's basic premise, which is that selection pressure brought certain apes out of the trees and directly onto the savannahs. For Ellis, various predators and competitor species of baboons were a significant obstacle to pre-hominid or hominid occupation of the savannahs.

He notes that the *'savannah is occupied by specialist long-ranging and fast cursorial herbivores and carnivores. Food would have been hard to get there, hard to keep, and getting it would be dangerous enough for hunting parties let alone camp followers comprising elders, pregnant females, and infants. In the process of adapting to parkland, not only the adult males, but also females and infants, would have to be viably adapted to the complexities of whatever ecosystems they were inhabiting at any one time (especially to predation) during the long series of anatomical, physiological and behavioural changes on the way to becoming human. And they would have to be able to push their way into the new ecological niche by adaptiveness superior to that of the baboons which had 'come down from the trees'* [Ellis' quotes] *before them. Therefore the evolutionary challenge was to transfer certain apes from a lower strata forest niche onto tropical parkland or riverine treed grassland, not the the open savannah'.* In any case, Ellis argues that the savannahs offer insufficient isolation between pre-hominids and apes to allow only the former to become bipedal.

Like several others, Ellis sees the geological events during the early Pliocene that formed the Rift Valley as fundamental to the climatic and ecological circumstances of hominid evolution. He proposes that hominid evolution was more plausible in marine rather than freshwater wetlands because marine wetlands are richer in immobile and slow moving edible biomass (food) and are also safer from predators than freshwater wetlands.

Like Lalumiere, Ellis sees the marine coast at the northern end of the Rift Valley near the Red Sea and Danakil Depression as a likely region of hominid evolution, but also thinks the southern end of the valley at the Zambezi River is a theoretically viable possibility. Both the oceanic rifting that created coastal archipelagos and volcanic activity in the Rift Valley would have destroyed many previously habitable areas. Bipedalism providing mobility to seek new habitats would therefore have been advantageous and might explain the widespread and scattered distribution of hominid fossil sites in Africa.

Ranked from wettest to driest, Ellis classifies marine wetlands as salt marshes, mangrove swamps, lagoons, rock shores, surf beaches, and sand dunes. Salt marshes support numerous edible plants, animals, fish, ground-nesting birds, worms and insects. They increase in salinity on the seaward side where molluscs, snails and aquatic crustacea prevail, and where fish occupy pools and streams especially at high tide. In the tropics, marshes also support trees particularly mangroves.

Lagoons are marine lakes that also vary in salinity but are generally protected from direct wind and waves. Lagoons host many grasses, invertebrates, fish, migratory birds, and aquatic herbivores. Rock shores host invertebrates, intertidal plants like algae and kelp, and birds and mammals that scavenge at low tide. Sand beaches and dunes are less productive either because of low rainfall, low ground water or continuous wave and tide action, but can still sustain dormant plants and oases. These various forms of wetlands can be adjacent to each other in any combination and are found in all types of climate the world over. Of particular relevance to hominid evolution, they exist from the Red Sea all the way down Africa's east coast.

In the time frame of hominid evolution, Ellis proposes that the Red Sea may well have been an isolated inland sea with variations in habitat and climate occurring both seasonally and due to periodic uplifting or subsidence caused by faulting. Higher temperatures and the dryest conditions would have confined the most productive wetlands to islands, coral reefs off shore, or oases inland. Cooler and wetter conditions would have expanded these wetlands and made many of the mangrove forests continuous with the rainforests inland. Whether isolated along the coasts (as in Borneo today) or occupying larger areas inland, these rainforests could have sustained a lower canopy or ground dwelling ape during the initial formation of the Rift Valley upwards of twenty million years ago or during the more active rifting about five million years ago.

If seasonal flooding of the Rift Valley lakes and river valleys occurred as it does today in the Amazon, a rich flora and fauna seasonally adapted to flood conditions would have developed. Ellis suggests that, in short order, improved knowledge both of the predators and necessary feeding cycles caused by predators or ocean tides would bring further adaptation. Some pre-hominid or hominid clades might well have inhabited islands and been relatively secure from predators. A variety of simple tools would have been used to crack open shellfish or crustacea or to slice open beached animals.

Tolerance of water would begin in children, which were developing fat and could easily float, especially in saline conditions. He notes that there is no implicit biological problem with a woodland ape evolving into a fully aquatic hominid, which, with speech, better dexterity, and higher cognitive function, would have adaptive advantages over the versatile baboon. The problem is in subsequently turning the aquatic hominid into a terrestrial human such that the original aquatic hominid becomes extinct.

Thus, geological events in paleolithic East Africa creating ecological settings suitable for a semi-terrestrial ape to become a bipedal wetlands ape are crucial for plausibility of a shore-based phase in human evolution. The real plausibility of such a habitat gives substance to the physiological or morphological features of humans discussed that are supposedly adaptations to shore-based or aquatic life. Hence, Hardy, Morgan, Verhaegen and other aquatic supporters cannot be dismissed on the grounds of lack of plausibility. Nevertheless, plausibility isn't direct evidence that a pre-hominid ape actually occupied, indeed depended on, such a habitat to become a hominid. Without evidence coming directly from hominid fossil sites, compelling support for wetlands or shorelines as a speciating force in human evolution would still be elusive.

Fish in the Hominid Fossil Record

Ten years ago, Kathy Stewart of the Museum of Nature in Ottawa published a paper in the *Journal of Human Evolution* documenting that

H. habilis in the Rift Valley purposefully exploited significant amounts of fish for food. Stewart is well known in the paleoanthropology community for her skill at identifying individual fish species in hominid fossil sites. She points out that the presence of fish remains has been well known for over thirty years at several hominid fossil sites around Lake Turkana in Kenya, Senge in Zaire, and Olduvai in Tanzania. Stewart has worked at several of these locations, notably with Meave Leakey at Lake Turkana. Despite their abundance, human paleoanthropologists have paid scant attention to fish remains in the ecology and food supply of hominids.

Fish bones are also plentiful at early human sites along the Nile River, and in Zaire, Botswana, and Kenya. At the White Paintings Rock Shelter site in south Africa, catfish remains are spread throughout more than two meters of deposits (or *middens*). Unlike Olduvai Gorge, these early human sites are typically late Pleistocene, with those of the Nile River being dated to about 40,000 years ago, while those at the mouth of the Semliki River are about 25,000 years old. Barbed points recovered amongst these fish remains suggest the use of spears to catch fish.

Stewart's evidence of long periods of planned, purposeful exploitation of fish by early humans motivated her to investigate whether fish were also a food resource for at least some hominids. Some hominid fossil sites, like Olduvai Gorge, have particularly dense fish remains that greatly outnumber the mammalian fossils. She and others have identified remains of both catfish *(Clarias)* and cichlids (i.e. *Tilapia,* a fish similar to perch) from more than 4,000 fish elements found at two hominid sites in Olduvai Gorge. Catfish are a bottom-feeding inshore fish growing up to two meters in length. They are tolerant of a wide range of water conditions including low oxygen and alkalinity. Cichlids prefer fairly shallow water near shores and can also tolerate poor water conditions.

A lower density of fish remains usually indicates the fish accumulated after dying of natural causes, whereas a higher density of accumulation typically implies that hominids caught the fish and left their bones behind. Regardless of the cultural artifacts that are recovered, catfish predominate at all but one of the sites. Stewart suggests this is not because of technical skill or species preference but rather can be attributed to the saline and alkaline conditions of paleo-Lake Olduvai, which eliminate many aquatic flora and fauna.

Clear, straight cutmarks on animal or fish fossils are unlikely to occur from natural causes or from animal predation, so they are a hallmark of hominid or pre-human butchery and consumption. In the absence of such direct evidence of handling by hominids, paleoanthropologists are reluctant to attribute the proximity of fish and hominid fossils as indicating active and purposeful exploitation of fish for food.

Catastrophic environmental events completely unrelated to hominid use can cause unusual fish bone accumulation, including wind storms or large amounts of run-off water. Successive deposits of fish bones are also more likely to occur with seasonally receding waters repeated over many years. As commonly occurs in modern times, stranded fish may or may not then be preyed upon by carnivores or other animals including hominids. Hence, the presence of fish bones amongst hominid fossils is encouraging but still inadequate evidence of purposeful or large scale exploitation of fish as a food resource.

Fish remains at several hominid sites recur in successive layers. The abundance of these remains and their distribution show that certain fish species were caught during specific seasons. At some sites, Stewart has identified as many as eight fish genera, with cichlids, if they are present at all, coming a distant second to catfish. In fact, catfish may represent close to 100% of the fish at most of these sites. As is the practice today, catfish and cichlids were probably caught during two seasons - on the floodplains of rivers during the wetter, spawning season and again during the dry season as waters receded.

On the basis of detailed assessment of the plant material that accompanies the fish remains, Stewart suggests that one of the richest sites of fish remains in Olduvai Gorge, known as FLKNN, was a lake margin site about one kilometer distant from the lake itself. Like other lakes in the Rift Valley, paleo-Lake Olduvai had no outlet so its water levels varied markedly with season and climate. Encampments would therefore be more likely at a site less prone to flooding and hence somewhat removed from the water's edge. Two levels of the FLKNN site contain very high densities of catfish remains. The size and age profiles of the bones imply that they were fished somewhat indiscriminately and unrelated to season or spawning. Level 3 of this site contains fossils of *H. habilis* and relevant cultural and animal remains whereas Level 2 contains only animal remains.

Stewart argues that several lines of evidence accompanying fish remains at Olduvai Gorge sites collectively represent purposeful hominid exploitation or predation of fish (Table 13.1). Uniequivocal cut marks are rare on fish bones but appear likely on several specimens she has studied. In addition, higher accumulation of fish heads rather than the vertebrae suggest the heads were detached prior to consuming the rest of the fish elsewhere. Natural causes of massive fish die-offs usually result in equal ratios of fish heads to backbones so successive deposits of fish remains containing skewed ratios of cranial and vertebral bones are unlikely to be attributable to environmental causes.

Table 13.1. The main features leading Kathy Stewart to conclude there was a high probability of active predation of fish by *H. habilis* and *H. erectus*. In some cases, these features are common to fish assemblages uninfluenced by hominids or other predators, i.e. they occur due to natural disasters. However, in her view, the joint occurrence of five or more of these characteristics in well perserved and relatively large fish assemblages at early Pleistocene fossil sites in Olduvai Gorge provides good evidence of intentional hominid exploitation.

1. The fish fossil sites are located at the banks or mouths of rivers.
2. Few species are represented.
3. Evidence of selective exploitation occurs according to seasoning and spawning habitat.
4. The fish remains are from species that can be caught with minimal technology.
5. There is a skewed distribution of skull and vertebral fragments in the deposits.
6. Several sites show evidence of repeated hominid occupation.
7. In a few cases, there is evidence that hominids probably made cutmarks on the fish bones.

Stewart also notes that the fish skulls in particular are fragmented in the hominid fossil sites. Natural die-offs generally cause little fragmentation of fish bones, especially the skull but crushing the fish skull prior to eating its brain is a common practice in present-day African groups. The hominid fossil sites in the Olduvai Gorge also contain turtle shells without other remains of the turtle which also indicates probable hominid predation.

Despite the presumed consumption of meat by *H. habilis*, Stewart observes that present-day hunter-gatherers in Tanzania and South Africa have a relatively poor record of success capturing medium to large mammals. During study periods averaging six to nine months, the kill rate by two such hunter-gatherer groups was about one such animal every three to six days. As the sole source of food, this success rate is insufficient to feed their typical group sizes, which number around fifty. During seasonal climate extremes, all mammalian sources of protein become harder to obtain and are often extremely lean, indicating other food sources are needed.

Stewart notes that fish are a staple and nutritious food resource for many present day African groups that still maintain a traditional lifestyle. Fish are equivalent to meat in protein quality and, seasonally, are commonly richer in fat than animals. As documented in Part 1 of this book, their mineral content, especially brain selective minerals, is unmatched by almost all terrestrial food sources, plant or animal.

Several indigenous African and Australian groups are well-documented to seasonally select certain species of fish particularly for their fat content, and will reject those that are too thin as being in 'poor condition'. Increased fatness in fish is commonly associated with the dry period just before spawning. Fish caught during the dry season would therefore easily replace the reduced availability and quality of many plants and animals.

Successful 'low technology' fishing, i.e. by hand or with spears or clubs but without lines or hooks, requires a knowledge of the relation between fish type, spawning period, habitat, and season. In east Africa, the main rainy season is March to April and may intermittently recur in November. Catfish are best caught in the wet season as they move upriver before spawning. They may also be caught returning downstream after spawning because the onset of the dry season strands many of them in shallow pools as the waters recede. With such knowledge, strategic planning can yield plenty of fish caught by hand. The addition of a little crude technology, such as sharpened sticks or clubs makes the process even more efficient, but is unnecessary if one is in the shallow waters commonly preferred for spawning. Catfish are abundantly caught in this manner, as are some cichlids.

Given the seasonal and short term nature of catfish or cichlid availability, this exploitation would have been planned and might well have involved competition from carnivores or other predators. Still, large groups of individuals could be fed this way. All age groups could also participate because fish stranded in shallow water are easy to catch.

Several present day African groups including the Dinka have seasonal festivals based on such low tech fishing. Procuring stranded fish from lake margins or pools would have been less challenging than catching them during spawning migrations on rivers. Peak fish availability in the dry season contrasts with that of most other food resources, making fish exploitation by hominids more plausible even in the face of potential competitors including carnivores, hyenas, and certain primates like baboons.

What does confirmed consumption of fish establish about the lifestyle or habitat of pre-human hominids two million years ago? First, it puts some of them on the margins or shores of lakes and rivers during at least part of the year. Knowledge of fish migration or stranding and careful selection of fish in 'good condition' (as some African fishers do today) requires a presence on the shores and some attention to be paid to what goes on in the water. As Stewart speculates, low technology fishing can be a group activity that, unlike hunting, is inclusive of the young, the pregnant, the nursing, and the elderly. Being on the shores for extended periods provides the opportunity to exploit other shore-based food resources besides fish, including shellfish, turtles, birds, shore and shallow water plants, etc.

If we accept that *H. habilis* was fishing, this necessitates at least wading in shallow water. Thus an opportunistically bipedal hominid would be likely to improve stance and gait while wading in the water to fish or look for other food. Attempts at floating, paddling, swimming and head submersion leading to diving would occur opportunistically but more frequently with habituation.

Better than almost any other food resources, fish consumption in *H. habilis* and *H. erectus* would contribute towards the metabolic and nutritional requirements of brain expansion. Brain expansion and sedentary encampments would catalyse social activity, speech, play and cultural activities. Tool making and art would be natural sequelae. The trigger in this scenario may well have been *need*, i.e. the seasonal need to find alternative food, but it wasn't the Savannah Theory's presumed need to make stone tools and hunt in order to find more food; everything about shore-based food procurement is low tech.

Intentional fishing by *H. habilis* doesn't adequately explain the origins of bipedalism or the gradual, early hominid expansion of the brain. These two features of *Homo* remain hard to explain unless fishing (or its nutritional equivalent) was occurring at least four million years ago. Stewart notes that the evidence for organised, even intensive, fishing by *H. habilis* couldn't arise overnight, so she feels that in some form or other it must have started in Australopithecines. How much earlier is unknown.

Despite the plausible geology and ecology, and the speculations of Hardy, Morgan, Crawford, Broadhurst, and Verhaegen, there simply isn't any firm objective evidence for early Australopithecine exploitation of shore-based foods at the level found by Kathy Stewart for *H. habilis*. Not yet, anyway. That may change if she has a chance to investigate the early hominid sites in northern Ethiopia.

Shellfish

Robert Walter is presently at the Department of Geology of the Centre for Scientific Investigation and Higher Education at Ensenada, Mexico and was previously in the Department of Anthropology at the University of Toronto. He has described clear evidence of exploitation of shellfish by early humans living on a coral reef on a bay on the present day Eritrea shores of the Red Sea. His group identified shells of two species of oysters, 31 species of molluscs and at least one species of large, edible crab in these coral reef and limestone deposits dated to 125,000 years ago. Sharp-edged tools were found in abundance, usually amongst the shell middens in and around large oyster beds. Walter concluded that the flake stone tools were used to harvest and crack open the shellfish or crustaceans.

Walter's report is consistent with others describing early human or pre-human exploitation of coastal shellfish but his group's date of 125,000 years ago is the earliest firm date for significant consumption of such a shore-based food resource. They found no pre-human or human fossils associated with these shellfish remains but contemporaneous African hominid fossils are clearly all early or fully modern *Homo sapiens*.

Walter suggests that dry inter-glacial conditions at the time would have encouraged exploration of coastal food resources, as would the relative lack of competition or predators on the coasts. He speculates that the extensive exploitation of the Red Sea site his group discovered implies that the first appearance of coastal food procurement would have been much earlier than the 125,000 year old date of their samples. Walter's speculation repeats Stewart's conclusion that if some *H. habilis* were regularly consuming lots of fish, some Australopithecines must also have been familiar with fish or shellfish as food.

Walter may not be prepared to say that exploitation of coastal shellfish by early humans more than 125,000 years ago implies that *H. habilis* was eating shellfish about two million years earlier. He would be even less likely to speculate about Australopithecine consumption of shellfish. Still, like fish in shallow water, molluscs and crustacea are shore-based foods requiring no specific skills or knowledge to find and eat. On some shores their abundance is overwhelmingly obvious. Though the fossil evidence is not yet available (and may never be conclusive), as Stewart speculates, if hominids before *H. habilis* were eating fish, they were probably also eating shellfish. Probability isn't proof but several pieces of the puzzle fit with a preliminary conclusion that some Australopithecines were eating both fish and shellfish.

Whether the shellfish consumed by pre-human hominids were on the sea coasts or were strictly freshwater is moot. Before and during the Pleistocene, the paleo lakes of East and North East Africa are known to have been variously saline and alkaline, often extremely so. Their input or drainage was often isolated and may not ever have been connected to the sea. Climate conditions varied considerably, both by season and over multi-year periods.

Some lakes and rivers systems were connected via the Rift Valley to the sea while others were not. A variety of fish, molluscs, crustaceans and other shore-based life, both plant and animal, have been conclusively documented in widely dispersed hominid fossil sites by Stewart, Walter and others. John Parkington and colleagues have similar data from large shellfish middens in early human fossil sites in South Africa but most of their material is dated later than Walter's.

The point is therefore to establish not that a specific type of food was consumed but, rather, that a broad and variable habitat zone on or near the shores was being exploited, first by certain groups of early hominids and then almost continuously by early humans. This exploitation would have been climatically forced in some areas or at some times and would have been optional at others. Either way, the use of shore-based foods, above ground or in shallow water, would have facilitated habitual bipedalism and gradual brain enlargement in successive groups of hominids. Hominids that did not continue to exploit this rich ecological zone would not have become humans.

Isotope Archeology

The emerging field of *isotope archeology* has implications for determining food resources used by hominids. Different numbers of neutrons change the weight of the atomic nucleus creating different forms *(isotopes)* of individual elements. Isotopes can be unstable (radioactive) or stable (non-radioactive) but they still represent the same element. For instance, there are three isotopes of carbon with atomic weights of 12 (carbon-12;), 13 (carbon-13; ^{13}C) or 14 (carbon-14; ^{14}C). The main form of carbon is ^{12}C which represents almost 99% of all carbon in the earth and in all organisms. ^{13}C is carbon's non radioactive (stable) isotope and represents about 1.1% of all carbon. ^{14}C is the radioactive and quantitatively very minor third isotope of carbon.

The relative amount of the stable isotope of an element is commonly presented as a ratio to that of the principal isotope, i.e. the ^{13}C to ^{12}C ratio. The analytical method of determining the ratio between two isotopes of the same element is called *isotope ratio mass spectrometry*. Differences in the isotope ratio of an element can be very informative about the chemical or biological origin of compounds containing that element. In turn, some isotope ratios potentially indicate the food origin of molecules in a living or fossil organism. Importantly, because the isotopes themselves and their ratios are stable, this methodology is applicable to the study of very old samples.

The ^{13}C to ^{12}C ratio is one of several such stable isotope fingerprints useful in paleoanthropology. In organic molecules, such as those remaining in animal or hominid fossils, this ratio varies depending on the plant source that first provided the carbon. The ^{13}C to ^{12}C ratio in different plants falls into one of two different ranges because plants have two different ways of processing carbon from carbon dioxide to make their organic molecules. Some plants use a three carbon intermediate between carbon dioxide and organic molecules. These are called the C_3 plants, which include the large majority of wild and domesticated plants worldwide, including most bushes, trees and shrubs. Other plants, particularly tropical grasses eaten by grazing herbiovores in areas with low rainfall, use a four carbon intermediate, so they are called the C_4 plants. Corn and millet are two model C_4 plants important in human paleoanthropology.

A *lower* ^{13}C to ^{12}C ratio is typical of the C_3 plants and the browsing, leaf- or fruit-eating animals that that eat them. A *higher* ^{13}C to ^{12}C ratio is characteristic of the C_4 plants and the grazing animals that eat them. The ^{13}C to ^{12}C ratio has been widely used to evaluate when pastoral or coastal peoples in Europe started to switch to corn- or millet-based agriculture. Placed in the context of sound cultural and fossil artifacts, the ^{13}C to ^{12}C ratio is an excellent index of the domestication of two agricultural plants - corn in the Americas and millet in Asia.

The ^{13}C to ^{12}C ratio also provides a promising new way to investigate whether shore-based foods were fundamentally part of hominid evolution and, if so, when. Starting with plankton, marine plants and subsequently the fish and marine animals that eat them have a higher ^{13}C to ^{12}C ratio than do most edible terrestrial (C_3) plants. In the absence of signs of cultivation of corn or millet, a higher ^{13}C to ^{12}C ratio therefore generally stands out as uniquely marking the consumption of food that must have been of marine origin, i.e fish or shellfish procured along the seacoasts.

Judith Sealy and Nik van der Merwe of the University of Cape Town, South Africa, have extensively investigated the contribution of marine foods to the intake of humans living on the coasts of South Africa in the past 10,000 years. Their stable isotope data using ^{13}C to ^{12}C and isotopes of nitrogen (^{15}N to ^{14}N) unequivocally confirm that the vast deposits (or middens) of shells at these sites occurred because early humans ate substantial amounts of shellfish such as mussels. Thus, these middens accumulated directly because of human consumption and not because of natural disasters or other causes.

Also using ^{13}C to ^{12}C and ^{15}N to ^{14}N ratios, Erik Trinhaus and colleagues from Washington University in St. Louis have reported that 20,000 to 28,000 years ago, early modern humans were consuming some fish and other aquatic food resources including molluscs and shore birds. Their analysis of Neanderthal specimens confirm what so many cultural artifacts have long suggested which is that most Neandertals were eating terrestrial herbivores. Just as the coaccumulation of shellfish middens with hominid or human fossils doesn't mean the shellfish were being eaten, so the animal bones in Neanderthal fossil sites don't prove these animals were being eaten. However, stable isotope ratios that are similar in the shellfish middens and animal fossils as in pre-human remains would make the direct food connection essentially irrefutable.

It is always satisfying when there are relatively few plausible options and the isotope ratios confirm the abundant cultural evidence at the fossil site. Sometimes, however, the fossil evidence is far from clear. Apart from a hominid fossil tooth or some other identifying skeletal fragments, there is little supporting evidence of any kind for what diets hominids were consuming. Julia Lee-Thorp from the Department of Archeology at the University of Cape Town, South Africa, has dealt with this situation. Her group has used stable isotope methodology in several investigations to assess the diets of much earlier hominids such as *A. africanus*, but she has still been forced to guess the implications of certain results.

For instance, Lee-Thorp showed that the ^{13}C to ^{12}C ratio in tooth enamel from Australopithecines unearthed in Makapansgat, South Africa, was somewhat higher than in C_3 plants but not as high as in truly mixed consumers of C_3 and C_4 plants. The large, thick molars of Australopithecines also imply that the moderately elevated ^{13}C to ^{12}C ratio arose in part from grinding fibrous, tough C_4 plants, i.e. tropical grasses. Her values in these early hominid specimens were almost identical with those in hyenas and quite different from those of committed vegetarian primates that rely mostly on C_3 plants. This led Lee-Thorp to conclude that *A. africanus* had a varied diet and was consuming a few C_4 but mostly C_3 plants, or was consuming meat from the animals that ate such plants.

The higher ^{13}C to ^{12}C ratio in Lee-Thorp's Australopithecine specimens could also have come from consuming shore-based foods, i.e from lake margins or rivers, but she didn't address this possibility. As Kathy Stewart speculates, some *A. africanus* or equivalent Australopithecines, were probably familiar with shore-based foods. Perhaps those around Makapansgat were not eating fish but there is general agreement that Australopithecines were opportunistic feeders. Lee-Thorp's stable isotope data are ambiguous about what foods *A. africanus* was consuming in South Africa and do not exclude shore-based foods. Where climate conditions permitted it is now becoming clear that their opportunism would have brought some Australopithecines face to face with shore-based foods including fish, shellfish, birds, amphibians and plants.

Other stable isotope ratios and ratios between different elements may also be helpful in evaluating the intake of shore-based foods, particularly those on marine shores. Nitrogen ratios (^{15}N to ^{14}N) in fossil material decline slightly but significantly as intake of plant material increases. The ratio of barium to strontium goes the other way – it decreases from a high in terrestrial sources to middle values in freshwater specimens to a low in marine species.

Dental enamel from teeth is laid down earlier than bone and, if remodelled at all, is changed much more slowly than bone. Hence, differences in isotope ratios between tooth enamel and bone of adults can indicate whether diet changed over the lifetime. Tooth enamel is also much better preserved in animal fossil material from the ancient past than is bone collagen so it is a vital source of material for stable isotope study.

Still, as Henry Schwarcz of McMaster University in Hamilton, Ontario, points out, there is no free lunch. There are significant methodological pitfalls in applying stable isotope methodology to evaluate fossils for evidence of dietary preferences. The first requirement – clean and uncontaminated starting material – is paramount and can be difficult to adhere to. Over time, inorganic elements infiltrate bone to fossilize it. This contamination is less likely with organic elements such as carbon but can occur if organisms like bacteria occupied any surfaces or porous matrix of the fossil.

Second, and equally importantly for hominid studies, 2 to 3 million years is a long period of time to preserve an organic molecule in bone. Most don't make it. Amongst the few organic molecules in bone that offer the greatest hope, collagen has been the most intensively explored. Third, isotope ratios vary for many reasons including climatic or geographic changes, etc., so the impact of these confounding effects needs to constantly be borne in mind.

The evidence for fish consumption by *H.habilis* obtained by Kathy Stewart is about as good as it gets. Nothing else puts hominids in such close contact with shore-based foods. Isotope ratios have the potential to clarify whether shore-based foods were consumed but have not yet been fully applied to this question. In some cases, isotope ratios will put the interpretation beyond reasonable doubt. In most cases, isotope or strontium-barium ratios will simply add to a weight of evidence suggesting a higher or lower probability of certain food options. That is still important because there are so few ways the problem can be studied in ancient specimens.

Iodine Measurements – Another Clue?

In addition to the isotope analyses of fossils that are known to be relevant to identifying food resources, iodine is another element that bears on the issue of shore-based foods in human brain evolution. Chapter 6 makes the case for iodine's importance in brain development. Chapter 14 will return to iodine via the thyroid and its potential role in speciation. Sebastiano Venturi of the Health Service in Pennabilli, Italy, points out that for all its importance in thyroid function, more iodine is located outside the thyroid than within it. He and his colleagues note that iodine is more abundant in the marine environment.

They propose that iodine may be an antioxidant that limited terrestrial vertebrate evolution until the thyroid evolved and was able to concentrate iodine for use in tissues other than the thyroid. In the absence of a thyroid gland *per se*, thyroxine still existed in the iodine–rich marine milieu where invertebrates evolved hundreds of millions of years ago. Venturi suggests that progression to the low iodine setting of terrestrial life required animals to develop a thyroid gland to increase thyroxine production.

Higher thyroxine could, in turn, be recruited as a means of getting more iodine into cells. The more active form of thyroid hormone, triiodothyronine, remained a hormone committed to controlling metamorphosis and thermogenesis but he speculates thyroxine really became an iodine transporter because it is more efficient than several well-known antioxidants including glutathione, vitamin E, and ascorbic acid.

Focussing Venturi's concept of iodine as an antioxidant on the need to protect vulnerable polyunsaturated fatty acids from lipid peroxidation, one could well see that further commitment of polyunsaturates like docoshexaenoic acid to brain membrane specialization in hominids would be curtailed in the absence of sufficient antioxidant protection. If any iodine-rich compounds survive over two million or more years, they would provide continuity between Stewart's evidence of hominid consumption of fish and the ongoing vulnerability of humans to iodine deficiency to the present day.

Chapter 14

How Would It Work?

We have examined the plausibility that conditions existed in East Africa that brought shore-based nutrition into play and could reasonably have contributed to human brain evolution. The issue remains as to how these nutritional catalysts could influence the genome in a way that created improvements in cognition that were transferred to future generations? Something has to mediate such a change.

Thyroid Hormone

One possibility is that the environmental change involves a change in the dietary intake or endogenous production of thyroid hormone (*thyroxine* and *triiodothyronine* combined). Thyroid hormone has two major functions in vertebrates: it regulates the rate of metabolism in cells and it is required for normal early development. The brain, heart and kidneys have a continuously high metabolic rate, day in and day out. These three organs must be well supplied with blood vessels to meet the high oxygen and nutrient needs. When these conditions are not met, these organs fail rapidly, as in a stroke knocking out brain function and causing paralysis, or impaired blood flow to the heart causing a heart attack.

In contrast, other tissues like skeletal muscle are active only transiently. In athletes, skeletal muscle is more active than in sedentary office workers taking the train or car to the office every day. The typical shortness of breath and muscle weakness experienced by people who are unfit indicates insufficient oxygen supply to the muscles. Energy consuming processes like blood circulation, cognitive function or high physical activity levels depend on efficient energy generation in the mitochondria which, in turn, depends on optimal thyroid function.

The number of thyroid hormone receptors per cell is a key determinant of the ability to increase or decrease metabolism or cell development in response to thyroid hormone. In the laboratory rat, brown fat and the pituitary have the highest concentration of thyroid hormone receptors. After brown fat and the pituitary, the density of thyroid hormone receptors is somewhat lower in liver and kidney and still lower in the cerebral cortex. Immune cells have very few thyroid hormone receptors while none have been detected in testes or spleen. The lack of thyroid hormone receptors makes the testes and spleen totally unresponsive to thyroid hormone. Interestingly, the adult brain is also unresponsive to thyroid hormone despite having large numbers of thyroid receptors. Apparently the responsiveness of the developing brain to thyroid hormone switches off as the brain matures even though the receptors remain into adulthood.

Unlike development of the brain, which depends heavily on adequate thyroid hormone levels, normal development and function of the heart and kidneys is relatively insensitive to thyroid hormone. Thus, the basic shape and function of the heart or kidney does not differ much across vertebrates despite wide differences in body form, locomotion, and cognitive capacity. Evolution seems somehow to have optimized the structure-function relationship of the vertebrate heart and kidney and, in so doing, gradually desensitized them to thyroid hormone. In contrast, body morphology and brain capacity are still open to evolutionary change and are still sensitive to environmental challenge and modification by thyroid hormone, especially during early development.

The flexibility inherent in brain and body morphology is both an advantage and a disadvantage. Flexibility is advantageous because it helps individual species adapt to different or changing environmental circumstances, i.e. to evolve. A species might be able to evolve wings or modify the brain in a certain environment but, so far at least, wide differences in locomotion or food selection have not required a radical new design of the basic blueprint of the vertebrate heart.

On the other hand, inter-species differences in physical form and in brain development are a disadvantage, at least during early development.

because they show that conditions adversely affecting thyroid hormone production, i.e. iodine deficiency, seriously compromise physical development and cognitive function. Animals possessing more sophisticated neurological development are therefore vulnerable to impaired thyroid function but the malleability of their brain and physical development leaves them better able to take advantage of conditions that increase thyroid hormone production.

The Pacemaker

Thyroid hormone has a key role in three systems in the body: reproduction, fetal and early post-natal development, and the response to stress. Thyroid hormone controls cell division and differentiation, thereby effectively controlling the timing and duration of development. It is not only the single most important regulator of normal development, including metamorphosis of the tadpole, but it also links many morphological, physiological and behavioural traits. Normal variations in thyroid hormone production lead to subtle differences in phenotype, i.e. hair colour, stress, response, etc. The thyroid appears to be central to translating environmental factors into physiological and developmental responses that lead to sufficient genetic variability to allow speciation to begin.

Blood thyroxine levels normally vary in a *pulsatile* manor. The length of time each pulse lasts varies across different species. For instance, blood thyroxine pulses last longer in humans than in domesticated animals such as cats or dogs. Maximum and minumum levels of thyroid hormone pulses change daily and with progress through developmental stages from larva to adulthood in different fish, birds and mammals.

Cyclicity and pulsatile secretion are well established features of thyroid hormone production, with a rhythm and pattern that are species-specific during early development. Even within a single species such as the dog, differences in blood thyroxine level are sufficient to affect the receptivity to mating of some but not other dog strains.

There are also marked differences in thyroxine's *half-life* (length of time spent in the blood) between species whose young develop quickly or slowly. These rhythms and the modulator effect of thyroid hormone on secretion of hormones by the pineal, hypothalamus, pituitary, adrenal and gonads, give the thyroid a key *pacemaker* role exerted by no other organ.

Heterochrony and the Antecedents of Speciation

Heterochrony is the natural variation in the rate and timing of pre- and post-natal development that leads to differences in size and shape of individuals within a species. Susan Crockford is a developmental biologist in Victoria, British Columbia, who feels that heterochrony is responsible for much of the inherent variability in phenotype that is essential for natural selection. Indeed, Crockford and others believe that heterochrony is the most common and significant biological mechanism underlying speciation.

Differences in thyroxine's half life appear to be necessary to make heterochrony possible. Small differences in thyroid rhythms create variability in traits between individuals and also account for generation-to-generation differences in traits. Both of these attributes present the opportunity for phenotypic change, including planned change such as human domestication of animal species. The dynamics of domestication are particularly informative about the early events leading to speciation. Similar thyroid hormone responses in genetically distinct populations help explain similar physical, behavioural and life history traits in genetically distinct populations, i.e. convergent evolution.

In addition to the genetic control that creates distinct thyroid rhythms, thyroxine secretion changes in a species-specific manner according to age, season, reproductive stage, presence of disease and stress response. As a pacemaker hormone, thyroxine coordinates the adaptive response of the body to short and long term environmental changes. If environmental changes are rapid or severe, i.e. increasing heat or cold, reduction in water supply, increased predators, etc., this invokes a further level of reaction or stress response.

Environmental challenges, including novel habitats, are better accepted by stress-tolerant individuals. Stress tolerant individuals cope better with evasion of predators, searching for more tolerable temperatures, or scarce water, etc. Changing habitat or circumstances may even be attractive to such stress-tolerant individuals. Stress tolerance varies across individuals and can account for up to 20% of a population. Those who are immediately capable of adapting to the stress of a new environment become morphologically, physiological and behaviourally, i.e. phenotypically, distinct, yet still retain a close genetic relationship to the ancestral population. This distinction is emphasized by increasing behavioural and reproductive isolation yet is genetically almost invisible.

When forced to opt for or against for a novel habitat, stress intolerant individuals do their best to avoid the new stress, i.e. they simply stay home. Crockford proposes that, within any species, stress tolerant individuals become *founder populations*. Founder populations contain a subset of individuals with smaller variation in their thyroid hormone receptors than were that present in the ancestral population. Matings within such smaller groupings would be expected to result in retention and promotion of the distinctive characteristics of - the founder population.

Protodomestication

In most families of higher organisms, clear evidence is missing for intermediate discernible forms in the fossil record that show the bridge between species. For instance, there are still large gaps between different hominids that leave some doubt as to which form led directly to humans. Despite intensive ongoing expeditions to uncover hominid fossils, these gaps remain because of the relative rarity of hominid fossils. The gaps are usually interpreted as indicating that the morphological changes associated with speciation are very gradual and require a long time period to occur.

Crockford suggests that this is not necessarily true. She points out that some new species of fish have arisen within the past 200 years, a period of time that is evolutionarily instantaneous. Furthermore, the morphological changes that accompany domestication or even *protodomestication* in species like the dog can involve significant changes to the skeleton that are not nearly as gradual as is generally assumed. Hence, the absence of evidence of intermediate forms in the archeological record is not sufficient justification for supposing that speciation always has to proceed slowly or by extremely small increments.

One example of a rapid change in a characteristic that appears to lead to speciation is the inheritance of stress tolerance during the domestication of foxes on Russian fur farms studied by Belyaev starting in 1959. This domestication permitted the furs to be harvested with less effort and with less damage to the animals than would be caused by hunting and trapping, thereby improving their value. By selecting the foxes most tolerant of the stress of being near humans, remarkably high tolerance was developed within twenty fox generations.

Domestication of foxes had the fortuitous and unanticipated effect of selecting for animals capable of bearing two litters per year compared to the usual single annual litter, an effect that not only halved the time to accommodate these changes but was also clearly economically desirable. Coat coloration also changed and piebald patterns emerged within the same twenty generations. The domesticated foxes had smaller adrenals and higher fertility, and some started barking like dogs.

Hence, the first step leading to speciation is the initiation of a separate founder population within the parent population as a whole. The difference in the founder population typically includes higher stress tolerance. In these foxes, it was tolerance of humans; in another species, it might be tolerance to a different stressor such as drought, parasites or predators. Stress tolerance would be advantageous to any species facing difficult choices affecting survival.

Proto-domestication of foxes shows that part of the genome controlling expression of physiological responses can, within a few generations, rapidly accommodate changes in an innate response, i.e. to fear of humans. In the fox, the change in stress response was soon accompanied by changes affecting reproductive rate in the stress-tolerant founder population, which, in turn affected the rate at which the new variant expanded. Thus, expression of a gene or genes influencing thyroid function can be modified by the environment. This modification is fairly easily transferred to succeeding generations without requiring gene mutations *per se*.

Iodine is needed for optimal thyroid function, so it seems reasonable that geographical areas providing more iodine would improve the ability of the thyroid to mastermind developmental change in susceptible organ systems (brain and skeleton) as environmental conditions change. This rationale predicts that species adapt to a certain level of iodine intake and have a tendency to remain evolutionarily static unless iodine or thyroid function changes. Conversely, access to a diet providing higher iodine predicts change in thyroid rhythms, stress tolerance, etc.

Thyroid Hormone and Hominid Evolution

For normal development in humans, no organ has a higher dependence on thyroid hormone or on iodine than the brain. It therefore seems inevitable that dietary iodine and thyroid function are closely linked to the unique changes in brain function that occurred as one branch of Australopithecines became the founder population for early *Homo* and eventually *H. sapiens*.

Improved stress resistance was probably an essential feature of exploration that occurred as *H. erectus* colonized temperate regions within and beyond Africa. The curiosity and fearlessness about unknown neighbouring localities that became large-scale exploration would have been strongly supported by the emergence of true bipedalism which, itself, would have arisen more easily in a habitat providing more thyroid hormone and/or iodine.

Bipedalism probably arose by heterochrony in a founder population of pre-Australopithecine primates that were normally quadrupedal. Many primates that lack the distinct and dedicated bipedal skeletal morphology found in bipedal hominids and humans still occasionally stand on two feet. As observed in the virtual domestication of foxes, only two things are required to exaggerate such a tendency from opportunistic to fully developed bipedalism : First is sufficient intrinsic variability in developmental morphology of the skeleton. Second is a slightly different rate or timing of skeletal development in the founder (bipedal prone) group. Different habitat and food sources would probably have been key stimuli, not only to opportunistic bipedalism, but also to a changing thyroid rhythm, which would have been necessary to promote the required morphological changes in the pelvis and spine.

A change in global distribution is often forced on a species by climate changes causing habitat change. Major global climate changes have periodically occurred and appear to have predisposed emergence of successive waves of change in the primate ancestor leading to humans. During one of these climate changes about 16 million years ago in the Miocene epoch, African apes expanded into Europe and Asia. Further climate change five million years ago in the Pliocene appears to have facilitated conversion of some quadrupedal pre-Australopithecine primates into bipedal Australopithecines. The start of the Pleistocene epoch about two million years ago saw futher climate change and the emergence of *H. erectus*, who was the first hominid to leave Africa about one million years ago. Certainly, Australopithecines appear not to have left Africa.

Changes in dentition between Australopithecines and other primates indicate some Australopithecines were exposed to dietary diversification and broader selection of foods. Crockford suggests that these new foods probably included increasing amounts of insects, grubs, eggs, and fledgling or moulting birds. I would add shore-based animals like ground-nesting water fowl, fish that were relatively easily caught by hand (catfish), crayfish, shellfish, and some small reptiles or amphibians. Although some of these animals were perhaps already somewhat familiar food, none of them would previously have been consumed in any great amount because, initially, Australopithecines were neither forced to diversify their diet nor were they well equipped for it.

All animal tissues including eggs contain some thyroid hormone. They also contain more iodine than do plant-based foods (Table 6.1). Consumption of fish, shellfish, meat or eggs therefore inevitably led to intake, even if sporadic, of exogenous thyroid hormone. Small animals eaten whole (frogs, some fish, fledgling birds, etc.) would have given a larger surge of thyroid hormone because of consuming the thyroid itself.

Thyroid hormone is the only hormone that is absorbed intact during digestion. This increase in available thyroid hormone would have stimulated relatively rapid morphological changes in development of the infants of a founder population of hominids that were eating more meat, eggs and shellfish. At a minimum, the combination of higher intake of both exogenous thyroid hormone and iodine would have provided unmatched metabolic and developmental support for the skeletal changes and habits that were beginning to favour bipedalism. Bipedalism required better unconscious control by the brainstem and cerebellum. Improved manual dexterity, planning, and memory all involve improved conscious function of the forebrain or cerebrum.

Crockford notes that since thyroxine can be toxic in high amounts, a high enough maternal intake of thyroid hormone and or iodine would potentially cause maternal hyperthyroidism, which could reduce fertility and even produce malformations. Still, some malformations could have been favourable, e.g. pelvic adjustments that favoured bipedalism and birthing of a larger brained baby. If such morphological changes favoured survival and did not imperil successful reproduction, there was a higher chance they would be passed on to the next generation. Thus, to Crockford, it seems possible that functional (if not fully optimized) bipedalism could have evolved within a very short period, perhaps within a few tens of thousands of years.

If, as is widely portrayed, *H. habilis* became a skilled scavenger of animal carcasses killed by other predators on the grasslands or open woodlands of East Africa, they would have consumed thyroid hormone from the diet. Competing successfully with other well-adapted scavengers and grassland predators would have required substantial stress tolerance. Those *H. habilis* who chose the scavenging route to nourish themselves undoubtedly faced at least as demanding an environmental challenge as those who found themselves in areas providing minimally adequate availability of food and water. Ability to scavenge would certainly have benefited as much from improved stress tolerance as would those exploring other lifestyles and ecological niches.

Animal carcasses and other terrestrial foods consumed by the scavengers would have provided less iodine and pre-formed thyroid hormone than would have been in the fresh foods on shorelines. Hence, a land-locked clade of scavenging *H. habilis* would have undergone slower morphological change in response to environmental stimuli than would have occurred in those occupying the shorelines. As a result, improvements in cognitive function and dexterity of the land-locked early *Homo* would have occurred more slowly compared to those on the shorelines. This may at least in part account for the morphological mismatches that seem to occur in some Australopithecine and early *Homo* fossils, in which inordinately big brains occur in specimens with relatively primitive evidence of bipedalism, or vice versa.

Within 250,000 years of the emergence of *H. habilis*, *H. erectus* had started to evolve and probably co-existed with *H. habilis* for some period of time. The habitat of *H. erectus* grew rapidly to extend throughout the temperate regions of Europe and Asia. Fossil evidence in Polynesia unmistakably shows that *H. erectus* voluntarily traveled by water for at least 50 miles, a distance greatly beyond the horizon, i.e. beyond which destination land masses could be seen. Clearly, to have become familiar enough with water to be exploring blindly over such distances, some *H. erectus* were not only highly stress tolerant but were skilled enough to navigate water craft. Whether they intentionally fashioned these watercraft or were opportunistic 'boaters' is unknown.

The underlying point is their evident willingness to increasingly engage in novel and potentially stressful activities unfamiliar to previous generations. Unlike most mammals, some *H. erectus* clearly seem to have accommodated easily to water. This suggests that at least in some areas, *H. erectus* had lifelong familiarity with water; they must have lived near it, played in it and used it to explore new coastlines. The enjoyment of water by humans contrasts with the fact that most primates spend little or no time in the water, so some form of stress tolerance was needed to alter this situation.

Neanderthals – A High or Low Thyroid Variant?

Neanderthals colonized colder regions, so Crockford describes them as a cold tolerant variant of their predecessor, *H. heidelbergensis*. Cold tolerance is dependent on thyroid hormone production, so Neanderthals may have been a *high thyroid variant* of *H. heidelbergensis*. More thyroid hormone production could also help explain why Neanderthals were physically massive. They ate more meat because they had acquired sufficient hunting skills and because of the lower availability of edible plants (fruit, nuts, seeds) in northern latitudes with longer cold seasons. Still, the meat was mostly raw because Neanderthals left little or no evidence of cooking. This would have resulted in higher thyroid hormone intake than in *H. erectus*, and might account for faster growth rate of Neanderthal compared to human infants. Neanderthals may well have had their own thyroid rhythm distinctive from other pre-humans.

Nevertheless, other attributes suggest Neanderthal had inadequate access to iodine. Jerome Dobson of the Oak Ridge National Laboratory in Tennessee notes that several Neanderthal skeletal and morphological traits are observed in endemically hypothyroid, iodine deficient humans (*cretins*). They include muscular hypertrophy, disproportionately long trunk relative to the short arms and short (bowed) legs, large head, low forehead, flared nostrils, marked brow ridges, large, rounded eyes, high bone density and tendency to arthritis.

Dobson describes how early explanations of the bowed femurs of Neanderthals were variously, but erroneously, attributed to rickets or arthritis. Rather, he believes that Neanderthals were essentially cretins, characterized by iodine deficiency from a combination of inefficient iodine uptake and excessive distance from the iodine-rich foods available on coasts.

The Venus figurines of the Cro-Magnon of the Upper Paleolithic period in Europe resemble cretins in the marked curvature of the spine, protruding belly, retracted (inverted) nipples and visibility of female genitalia. Dobson suggests that these features were used by the artists and sculptors of the day to model their figurines not because of some stylized but distorted female image but because they were prevalent in many women, most of whom were iodine deficient.

Dobson suggests that periodic sea level fluctuations caused by continental glaciation 200,000 to 20,000 years ago increased food competition and population pressures inland. This competition forced some clans to look for food elsewhere, including on the coasts, where they inevitably, and fortuitously, found shellfish. Those who established themselves on the coasts consumed iodine-rich, shore-based foods and became humans. They were able to force the less competitive clans away from the coasts and further inland. This would have pushed them towards the caves and mountains nearer the cold glacier fronts where, generation after generation, they became progessively more iodine deficient and where the most classic Neanderthal morphology arose.

Those with somewhat better but still intermittant access to coastal food resources had less severe symptoms of iodine deficiency and became the *progressive* Neanderthals. Migration from coastal settings would also help explain the presence at hominid fossil sites of mollusc shells hundreds of kilometers from their native coastal habitat.

Dobson postulates that a mutation helped convert certain Neanderthals into the earliest Cro-Magnon stock. By making the uptake of iodine by the thyroid more efficient, this mutation helped overcome the nutritional and population pressures created by advancing and retreating shorelines and allowed both streamlining of the newly human torso as well as cognitive improvement of the already large brain. In effect, Dobson suggests that Neanderthals were a diseased version of Cro-Magnon. He suggests that a genetic mutation increasing the efficiency of iodine uptake by the thyroid released the Cro-Magnon from this metabolic stranglehold, particularly on brain function.

The attractive element in such a pathological explanation is that it accounts for several unexplained features including the probable co-existence of Neanderthals and Cro-Magnan, the limited cultural and hunting repertoire of the Neanderthals, and their very rapid disappearance during less than 10,000 years. Successive waves of migrants from healthy coastal locations would have contributed to maintaining clan density in certain areas as well as skill levels for both hunting and modest cultural advancement. The higher incidence of symptoms of iodine deficiency in women supports the fossil evidence suggesting female Nedanderthals were more sedentary and less mobile because of skeletal pathology and obesity. If Dobson is correct, Neanderthals are the evidence that iodine deficiency crippled and eventually killed off the last non-human branch of hominids.

One way the discrepancy might be resolved between Crockford's high thyroid/iodine and Dobson's low thyroid/iodine explanations of how thyroid or iodine limitations contributed to the rise and fall of the Neanderthals is that high protein intake contributes to iodine loss from the thyroid. Thus, as Crockford proposes, by eating more meat Neanderthals may well have had higher thyroid hormone intake. This would have contributed to better cold tolerance and morphological changes increasing their skeletal and muscle mass. However, high intake of meat may have helped deplete iodine, thereby curtailing brain development in Neanderthals and limiting their cultural and technical development. Iodine deficiency could then have contributed to the pot-bellied physique which occurs in present day cretins who can fully develop physically but are neurologically impaired. A variant of Neanderthals with more efficient thyroid uptake of iodine and/or better access to dietary iodine became the Cro-Magnon who noticed and sculpted the unique anatomical features of iodine deficient Neanderthals into their figurines before they became extinct.

Fetal Programming

The vulnerability of the brain is mostly within the individual. An iodine deficient person may be mentally retarded but, if fertile, can still produce a baby. If it can avoid iodine deficiency, that baby will have the potential for normal brain development. Environmentally-induced deficits (or advantages) in development that occur in the parent are not inherited by the next generation. However, as soon as the egg is fertilized, it becomes susceptible to the conditions of life. The blueprint for the frame of the house is laid out but the final form the house takes, i.e. whether it will have a leaky roof, air conditioning or elegant furnishings, is still unknown.

Translation of the DNA code into phenotype is sensitive to the conditions of embryonic and fetal development. Over the past twenty years, two separate research teams led by British scientists, David Barker of the University of Southampton and Alan Lucas, originally of University of Cambridge but now at the Institute for Child Studies in London, have independently developed this concept, which is called *fetal programming*.

Both groups found that environmental stresses causing a human fetus to develop a little slower or be born a little smaller could significantly raise the risk of death from heart disease. Babies born small had higher risk of premature diabetes, hypertension and death from heart disease. Smallness was more likely if your mother was small, smoked or had malnutrition. In effect, environmental conditions that impaired the function of the placenta or restricted fetal growth had long-term ramifications well into adulthood. Regardless of one's nutritional good fortune after birth, the enhanced risk remained.

Still, higher risk acquired during stressful fetal development can be attenuated by a healthier post-natal lifestyle, or can be exacerbated by an unhealthy lifestyle. Low body fat content is a big part of the problem with slow fetal development and low birth weight, but the reasons why this is so remain unclear.

Two important points emerge from Barker's and Lucas' observations: First, getting a full complement of defect-free software (DNA) doesn't guarantee that the hardware (the organs and body as a whole) will work perfectly. Translating the genome into all the essential functions constituting a complete organism is open to manipulation at various points starting during development of the embryo. Second, disadvantageous conditions during fetal life increase the risk of premature heart disease without changing DNA *per se*. Hence, in effect, a phenotype for premature heart disease is acquired before birth, not because of a defect in the DNA code itself but because of the way the code is interpreted as the fetal cardiovascular system is developing.

Genetic control of protein synthesis, enzyme activity, thyroid hormone levels, etc., is responsive to regulatory molecules controlling *gene expression*. Gene expression is the flexible aspect of interpreting DNA. It is where the environment affects development, disease risk and, in the case of human brain evolution, many aspects of brain maturation and function from neurogenesis to myelination to memory.

Barker's and Lucas' analysis show that if your mother was small and you maintain the same lifestyle as your mother (especially if you are a woman), you are more likely to be smaller in stature and to transfer that smallness to your baby. Your DNA complement remains unaffected but its translation is compromised. With careful attention to an optimal lifestyle, this situation can reverse at the same rate that it can deteriorate. In effect, the environment can impose on natural selection and on the inherited DNA an influence that, though uninherited *per se*, affects the second generation. Barker and Lucas have also documented fetal programming effects predisposing to higher cardiovascular disease risk, particularly those involving obesity, hypertension, insulin resistance and diabetes. Others, like Lucille Hurley's group at the University of

California-Davis showed twenty years ago that the immune system in mice likewise remains 'scarred' for multiple generations by malnutrition involving zinc deficiency.

Amongst humans, the second generation usually adopts the same habits and lifestyle as its parents (the first generation), so the same environmental influence on phenotype can be propagated to the second generation and beyond. Negative effects of an adverse environment are correctable because the genotype has not changed, but they persist as long as the next generation is raised under the same conditions as the previous generation. Natural selection determines your genotype but the environment is the master of its translation into phenotype.

Fetal programming is therefore an important example of non-genetic multi-generation change in function that affects a variety of organ systems including the brain, cardiovascular system, the skeleton and the immune system, in which the genotype is not changed but the phenotype is, even when the environmental insult is later lifted. This shows that in addition to changes in genotype, environmental influences can be transferred to the next generation. There is no (known) change in genotype, but neither is it Lamarckian inheritance. Rather, fetal programming is a gray area permitting long term environmental effects on succeeding generations.

Gene Defects in Nutrient Metabolism

Seriously impaired absorption or metabolism of a single nutrient is uncommon but genetic defects with this outcome do occur. Acrodermatitis enteropathica is a rare but devastating and rapidly fatal disease in which a single gene defect specifically impairs the intestinal uptake of zinc from the diet. The ensuing serious zinc deficiency impairs growth, protein synthesis and immune function.

Other examples include Wilson's Disease, which impairs copper excretion, and phenylketonuria impairing phenylalanine metabolism. In fact, the adverse impact of these genetic diseases has been overcome only in the past 50 years and only because ways of environmentally (nutritionally) managing them have been developed. Thus the negative effects of single gene mutations depend very much on environmental circumstances, particularly the intake of affected nutrients. Clearly, many gene mutations causing disease affect pathways unrelated to nutrient metabolism and cannot be corrected by a change in nutrition. Nevertheless, quite a few mutations do affect nutrient metabolism and can be largely overcome by altered nutrient intake.

Gene defects causing specific defects in nutrient metabolism are devastating but still rare diseases. On the other hand, *cystic fibrosis* is not only the most common inherited disease in Caucasian males but also causes more generalized malnutrition. The main problem in cystic fibrosis is with the function of the pancreas and intestines. As a result, absorption of fats and fat-soluble vitamins (but not other nutrients like minerals, sugars or proteins) is markedly impaired. In the long term, impaired lung function, a strained immune system and intestinal malabsorption reduce appetite and food intake and cause malnutrition.

When the first gene for cystic fibrosis was discovered in 1989, many dared to dream that this was the beginning of the end of cystic fibrosis. Now, however, over 1,000 different mutations in the cystic fibrosis gene have been discovered. Most of them are clinically silent (phenotypically normal). In fact, 70% of symptomatic cystic fibrosis cases have one of only a dozen different mutations. Scientists studying the molecular genetics of cystic fibrosis can now tell expectant parents whether their fetus has no gene for cystic fibrosis, an abnormal but silent gene or, in the worst case, one of the few versions of the dreaded abnormal gene actually causing cystic fibrosis. Initially, looking at the DNA sequence didn't indicate which was which unless the living person was available to compare the mutations to the presence or absence of the disease.

Cystic fibrosis is a good model of several important features of gene-nutrient interactions influencing human health: First, it is a relatively widespread defective gene in humans. Second, it shows how many silent defective versions of the same gene can exist. Third, it is a model of how much is still unknown about relating a disease (or phenotype) outcome to a gene defect. Fourth, its management depends heavily on effective nutritional intervention. Antibiotics are essential to the effective treatment of cystic fibrosis but focussed, intensive nutritional management also makes a big difference to the expected life span in cystic fibrosis.

Cystic fibrosis is an example of why linking human brain evolution to specific, advantageous gene changes is so challenging. How do we know which gene changes will be silent and which will be advantageous to brain function and whether they will depend on certain nutrients or other environmental variables? How many intended variants of 'smart' mice have not been heard about because the mutation was silent? Beneficial nutritional effects on cystic fibrosis take several years to become obvious; will 'smart' mice still be smarter when they age?

How do Nutrients Affect Gene Expression?

Establishing that gene mutations affecting nutrient metabolism can be nutritionally managed or altered demonstrates that nutrition and genetics are not two solitudes. Genes are neither independent of nor more important than the environment. Whether we are talking about response to a cold or evolution of the brain, Darwin's 'conditions of existence' involve a complex dialogue between the genome and the environment. But how does it work? What is the mechanism by which nutrients affect gene expression and, ultimately, evolution?

An old but still valid example is one place to get started. In the 1960s, Myron Winick at Columbia University was one of the first into the yet to be named field of *nutrigenomics*. Winick was trying to determine how indispensible amino acids affect growth and development. He had two varieties of bacteria, one that could make one of the indispensable amino acids, *histidine*, and one that could not. Since the variety that could not make histidine was in effect lacking an essential nutrient for its growth, it did not grow well unless histidine was supplied in the culture medium. However, when the variety lacking the ability to synthesize histidine was given histidine, it multiplied better than the bacteria that could synthesize histidine.

One might expect that the two varieties would become equivalent once histidine was given to those that couldn't make it. On the contrary, not being able to make histidine was advantageous as long as histidine was given in the culture medium. Apparently, it was more efficient for growth in these bacteria to be dependent on the diet for histidine than to be independent of dietary histidine by being able to make it. Winick's interpretation was that if the enzymes usually needed to make histidine synthesis are no longer necessary because histidine is present in the diet, more DNA and mRNA are available for other requirements of cell division and growth.

Eliminating instructions for one amino acid from the cramped space in the bacterial genome frees up that space for other useful information that, in Winick's example, seems to have conferred an advantage because the bacteria depending on diet for histidine grew faster than expected. Bacteria grow and replicate very quickly so they also evolve quickly. In effect, histidine became a dietary vitamin and nutrition became of direct importance in regulating the bacteria's gene expression and evolution. The trade-off is the dependence upon the environment, in this case, for a nutritional supply of histidine.

Then why not free up as much space as possible in the genome by depending on the diet for all nutrients? Organisms do indeed depend on diet for many nutrients. Some nutrients can be made by some organisms but not by others, i.e. plants can make the vitamins and other organic nutrients needed by animals. Other nutrients, like minerals, can't be made by plants, animals (or bacteria).

The trick is to depend on the diet only for substances (nutrients) found in good supply in your chosen diet. Don't evolve in a habitat that fully meets your nutrient needs and then move to a habitat no longer meeting those needs. This is what has happened with humans during the past century – we now consume diets that do not meet our needs for brain selective nutrients and enormous numbers of people are paying the consequences.

Lipids and Gene Expression

Polyunsaturates directly modify the expression of several genes controlling the enzymes of fat metabolism, tending to inhibit fatty acid synthesis and increase fatty acid oxidation. Effects of polyunsaturates on fat synthesis and oxidation probably have an impact on fetal fat development in humans but, at the moment, this is entirely unexplored. Polyunsaturates also control genes for glucose transport into cells. These are very broad ranging effects that operate at different points controlling the post-transcriptional production of mRNA and hence the amount of enzyme produced as well as its activity.

One question that arises concerning a change in gene expression that affects brain function is - are gene-directed changes in function possible without corresponding changes in metabolism? The brain is energetically expensive for its size so there is a crude positive relation between 'smartness' and energy requirement of the brain. Perhaps energy metabolism studies are now being undertaken in genetically modified strains of 'smart' mice. Do smart mice eat more? Are they leaner or is there no discernable impact on metabolism of faster learning and better memory?

This is a key issue in understanding human brain evolution. How much did human brain evolution, i.e. the evolution of 'smartness', depend on altering brain energetics? Could hominids have experienced improvements in intelligence of 50% without needing to eat more food or would even a 10% improvement demand a commensurate increase in energy intake?

It is probably naïve to expect a simple link between brain function and energy metabolism because, even in humans of widely varying intelligence, we are unlikely to ever show that the energy requirements of the smarter exceed those of the less smart. Although smartness varies across humans, humans all have larger brains that are more energetically expensive than in chimpanzees or in *H. habilis*.

At some point energetics seems to have been involved in brain expansion and increasing sophistication of neural circuitry but, at another point, it is not very relevant. Hence, combining the subject of the molecular genetics of 'smartness' in mice with the subject of energy metabolism should lead us towards a better understanding of how energetics may or may not influence the possible advantage of a 'smart gene'.

Equally, how does iron or iodine deficiency affect the performance of 'smart mice'? Are they more resistant to the functional decline imposed by iron deficiency, are they more susceptible, or neither? Modeling of this sort is quite feasible today and should provide a much needed perspective on the relevance of gene mutations or modified gene expression to human brain evolution.

Ultimately, changes in gene expression induced by changing levels of thyroxine, iodine, iron or other nutrients in one generation are able to change gene structure before reproduction and transfer of the genome to the next generation. Naturally, there is no one mechanism for controlling gene expression, even by a single nutrient. How histidine appears to work in Winick's humble example does not explain how it works in a different situation, let alone how polyunsaturates affect gene expression.

Indeed, as this new field of nutrigenomics emerges, there is a blizzard of new data accompanying the application of new methods that require verification in real life situations. The next decade should see significant advances in the understanding of the exact mechanisms by which gene expression can be permanently changed in the presence of chronically different nutrient intake.

Chapter 15

Survival of the Fattest

The human brain evolved because certain hominids stumbled serendipitously across solutions to two major constraints on primate brain size and function, one a metabolic constraint and the other a structural constraint. The unique cognitive potential of the adult human brain emerged as a direct consequence of evolving neonatal body fat as insurance against the *metabolic constraint* - the voracious fuel needs of the infant brain. Neonatal body fat improved the fuel supply to the brain by providing an alternative fuel to glucose in the form of ketone bodies. Ketone bodies magnified the potential for more sophisticated communication between neurons, but only because the habitat and diet permitting the development of body fat simultaneously provided a richer supply of brain selective nutrients. These nutrients, particularly docosahexaenoic acid, met the need for additional membrane complexity, which released the *structural constraint* on neuronal connectivity.

Of these two principal constraints, it is the first one – fuel insurance for the brain - which most clearly distinguishes humans from other primates. This book is therefore called *Survival of the Fattest* because the fattest infants had the most insurance and stood the best chance of becoming the smartest adults. Whatever best explains the evolution of fat babies also explains big, advanced brains in humans. A lot of physiology and biochemistry has to change to improve both brain function and fetal adiposity and, for the latter, there was no previous primate blueprint. To evolve fat babies, one is obliged to look to the shores because it was only there that the habitat and diet were sufficiently stable and abundant, while also providing a significantly improved supply of brain-selective nutrients.

Plausibility, Prediction, and Parsimony

The Shore-based Scenario has three important features of a valid theory – plausibility, prediction, and parsimony. The *plausibility* is that compared to all other sources of food, the shore-based food supply is richest in brain selective nutrients (Tables 6.1 and 7.1 to 7.6). The hominid and early human fossil record in several parts of Africa supports purposeful consumption of shore-based foods, particulary shellfish and catfish, dating back at least two million years.

The *prediction* is that a functional deficit in brain development will occur if human infants consume inadequate amounts of shore-based foods. More specifically, if brain selective nutrients contributed to gradually improving hominid brain function, then lower intake of these nutrients should have three important outcomes : 1) it should increase the risk of impaired brain development in infants, 2) it should increase the risk of impaired brain function later in adult life, and 3) it should have a significant negative impact on the evolution of hominids unable to sustain a liberal dietary supply of energy and brain selective nutrients.

The World Health Organisation's description of the widespread debilitating vulnerability to iodine and iron deficiency during childhood is compelling confirmation of the first prediction. Low accumulation of fat on the fetus in humans, whether through intrauterine growth retardation or prematurity, confers an increased risk of subnormal neurological development. With respect to the second prediction (impaired brain function later in life), many studies now indicate that Alzheimer's disease is more prevalent in populations consuming low amounts of fish and other shore-based nutrients. The third prediction is supported by the extinction of Neanderthals, in whom 'non-competitive' brain function accompanied large brain size in hominids. The reason for the demise of the large-brained Neanderthals is unknown but Jerome Dobson makes a good case for iodine deficiency having a role (Chapter 14).

Despite having somewhat larger brain size, subnormal brain function also occurs in autistic children today. Autism is clearly not a simple nutrient deficiency but added zinc and iron during childhood can improve the prognosis. Insufficient food availability does not seem to be a contributing factor to the defective brain function in either autism nor in Neanderthals, so the common assumption that improved dietary energy supply alone was sufficient to account for pre-human brain expansion is not well supported by these two examples.

Early hominids with somewhat less brainpower either watched and mimicked those that were slightly brighter who were already gathering fish and shellfish, or they competed with other primates in a woodland or savannah environment. If they were the former, they were more likely to become the *gracile* Australopithecines and to have the chance to evolve into *H. habilis*, *H. erectus* and, eventually, humans. If they were the latter, they were more likely to become the *robust* Australopithecines, who, by foregoing the abundance and availability of shore-based foods, unknowingly forfeited the accompanying benefits for the developing brain. Hence, added dietary energy might have helped get hominid brain expansion started but, without brain-selective nutrients, was insufficient to transform a hominid brain into a human brain.

Thusfar, other theories of human brain evolution have not shown this predictive ability : there is no evidence that low intake of alternative sources of dietary energy, such as meat or nuts, is associated with impaired brain function during either early development or during aging. Hence, diets that are not shore-based may have contained sufficient energy to meet the requirements of hominid brain expansion but they had and still have two serious inadequacies for human brain development : First, they are more likely to create nutrient deficiencies, particularly of docosahexanoic acid, iron and iodine. Second, plant-based diets contain *anti-nutrients* such as phytate and goiterogens, that exacerbate deficiencies of nutrients such as zinc and iodine, respectively.

The areas of the brain used for hearing need the most energy and are also the most vulnerable to dietary or congenital iodine deficiency. Good hearing is a key prerequisite for language, which is a defining feature of humans. Therefore, it is appropriate to think of iodine as a *sentinel nutrient* for human brain evolution because of its intimate link to the function of brain areas responsible for hearing and, indirectly, speech. Conversely, woodlands and grasslands occupied by non-human primates provided less iodine but these lower iodine levels were in keeping with lower demand for (and supply of) other brain selective nutrients. Lower iodine intake in fruit or vegetable-based diets wouldn't prevent other primates from having good hearing but would prevent a combination of both good hearing and brain expansion.

As John Langdon of Indianapolis University notes, *parsimony* is important in evolutionary theory; if a transformation during evolution can be completed in one stage instead of two, the simpler version should be favoured unless there is overwhelming evidence for a more complex process. The Aquatic Theory requires that the pre-human hominid lineage go through *two* adaptive phases. First it had to go from a semi-arboreal to a semi-aquatic habitat. Once transformed into a bipedal semi-aquatic ape, a second phase brought the bipedal ape back to terrestrial subsistence, at which point the semi-aquatic habitat was largely abandoned. Since some aboriginal groups seem well adapted to harsh inland settings where brain selective nutrients are clearly more limiting than on the shores, Langdon claims that the Aquatic Theory overcomplicates human evolution by adding a shore-based stage where none was needed.

On the other hand, the Savannah Theory proposes that the transition to bipedal human from quadrupedal ape involved only one fully terrestrial but necessarily more dramatic phase. The problem is that the Savannah Theory overlooks the suboptimal mental function and physical capacity of many hundreds of millions of iodine and iron deficient people, mostly in regions distant from the shores. Things may indeed be too complicated with the Aquatic Theory but they are also a bit too simple with the Savannah Theory. As Einstein noted, '*explanations should be made as simple as possible but no simpler*'.

Humans have clearly adapted to an incredible variety of ecological niches worldwide but it is a different matter as to whether they could actually have evolved in those difficult, challenging conditions where food resources can be stretched very thin. Once one is the possessor of a large and sophisticated brain and has access to centuries of training and experience passed down from previous generations, one is much better able to extract the necessities of life from an inhospitable environment. That is a much different thing than humans actually evolving from unspecialized hominids in such an inhospitable and competitive setting as the savannah.

The Shore-based Scenario is the middle ground between the Aquatic and Savannah Theories, both conceptually and ecologically. It provides not only the food and lifestyle necessary for human brain evolution, but also the geographic isolation that is generally more likely to foster speciation. Crucially, it respects parsimony because it involves only one phase – humans are *still in* the shore-based phase. Humans have explored other niches in the past and will continue to do so in the future but we are still best adapted to, and dependent on, a shore-based food supply.

Scars of Evolution

Support for the concept that we are best adapted to the shores is provided by a suite of diseases or chronic degenerative conditions to which humans are particularly vulnerable. Elaine Morgan describes these problems, particularly involving the joints and back, as *scars of evolution* (see Chapter 12). Leigh Broadhurst, Michael Crawford, Marc Verhaegen, Artemis Simopoulos, Loren Cordain and others add heightened susceptibility to several life-threatening degenerative diseases (atherosclerosis, hypertension, and insulin resistence) to Morgan's list of scars. A great deal of epidemiological evidence as well as numerous intervention trials demonstrate that humans consuming a shore-based diet are less vulnerable to these scars.

It is the ongoing developmental vulnerability of the infant brain which is the most significant scar of all. Humans didn't need to become divers or swimmers, nor did we need to become spear-wielding hunters to avoid these scars, but we did need moderately frequent access to a shore-based diet. The key trick to improving cognitive function during human evolution was to avoid exposing the brain's increasing developmental vulnerability as it expanded. This masking of developmental fragility was essential to refining brain function and was something that no other land-based species accomplished. Ongoing brain vulnerability during infancy makes it abundantly clear that humans have not yet distanced themselves from shore-based nutrition.

Improved problem-solving skills led to humans exploring and establishing themselves in inhospitable environments including the Arctic, the deserts, and remote inland or moutaineous regions. By discovering and legislating iodine supplements early in the 20th century, we have, in effect, modified the inhospitable low iodine environment inland. In so doing, the catastrophe of acutely impaired brain development has largely been avoided in many countries. Seven hundred million iodine deficient people worldwide attest to the fact that this success is still incomplete but that the failure is more political and logistical than scientific.

Environmental Permissiveness

There is great interest today in understanding the role of genes in human brain evolution. One important challenge facing these studies is to explain how genes that would have promoted increasing hominid brain size during a 2 to 3 million year period *stalled and then reversed* the increase thereby allowing the fairly rapid (although still small; see Chapter 2) decrease in human brain size during the past 30,000 years. Is it reasonable to postulate that gene-dependent processes account first for the increase, and then changed sufficiently to account for the subsequent decrease in brain size? In addition, is it reasonable to postulate that gene changes affecting brain size and sophistication coincided with those permitting additional body fat deposition so that brain expansion could be fuelled?

The environment or the diet can wax or wane; that is the nature of environmental events; sometimes environmental events make life more difficult but sometimes they can be permissive and make life easier in those who are genetically prepared to take advantage of the change. On the other hand, genetic information is more constitutive, so reversal of a 'beneficial' mutation is highly unlikely. Reversal of brain expansion in humans makes it implausible that genetic factors alone were the dominant force in propelling the changes in brain size and function leading to present day humans. Rather, the key to hominid brain expansion was *environmental permissiveness* (of the shore-based habitat), which still respects the requirements of natural selection and the role of genes but removes the pressure of survival as a speciating force.

Dolphins

Dolphins are perhaps the most cognitively advanced species after humans. Amongst the large mammals, their brain size relative to body weight comes closest to that of humans (1.0% compared to 2.2% in humans; see Table 2.1). Dolphins have some interesting behavioural capacities that until recently were believed to be uniquely human, including mirror self-recognition, the ability to understand artificial and symbol-based communication, and the inter-generational transfer of behaviours that can be equated with a form of culture.

Humans and dolphins have two other attributes in common besides large, cognitively advanced brains and certain behavioural similarities : they both have marked body fatness and a marine food supply. Dolphin body weight is 18-20% fat and up to 70% of the energy in their maternal milk is provided by fat. Their cold water, marine environment clearly demands a high energy intake in order to maintain body temperature but this high fat intake also serves to fuel the large developing brain. Dolphins subsist principally on fish and squid, both of which are rich sources of docosahexaenoic acid and brain selective minerals. The shore-based/marine food supply therefore seems to have been a crucial component in the evolution of the large brain and advanced cognitive capabilities common to both humans and dolphins.

Still, the dolphin brain has much more folding and a thinner 'neocortex' than in humans. It also evolved much earlier and under different circumstances than the human brain. Therefore, at best, one can say that the similarities between human and dolphin brain function represent convergent evolution. Nevertheless, docosahexaenoic acid is a prerequisite for normal brain function in all species in which the relationship has been studied so it is beyond reasonable doubt that dolphins would be cognitively advanced without consuming docosahexaenoic acid from fish.

Shore-Based Adaptations

Anna Gislen of the University of Lund has described the unmatched underwater visual acuity of the Moken 'sea gypsy' children of Thailand's Surin Islands who can gather shellfish at depths of 3 to 4 meters underwater without the aid of goggles. The Meriam children of the Island of Mer in the Torres Straits off Australia perform likewise. In contrast, untrained 'landlubbers' have poor visual acuity underwater and, without goggles, are unable to duplicate the skill of these children.

In fact, several present day traditional hunter-gatherers exploit shore-based food resources, sometimes in very inhospitable environments : The Inuit hunt seal, walrus and polar bear on Arctic ice floes. In northern Québec, they also go *underneath* the coastal ice to collect shellfish at low tide. The Ama of Korea and Japan, push the extremes of endurance to make a living by diving for undersea shellfish. Hence, exploitation of shore-based food resources is familiar to a wide variety of human populations in tropical as well as polar environments.

The origin of many human features such as relative hairlessness, bipedalism, and subcutaneous fat is contentious, whether explained by the Aquatic or Savannah Theory. In compiling the evidence supporting the Aquatic Theory of human evolution, Hardy, Morgan, Verhaegen, and others propose these features arose as adaptations of early hominids to an aquatic or semi-aquatic environment. But as Derek Ellis points out, the adaptations of aquatic and semi-aquatic mammals to their chosen habitat differ depending on the species. Some aquatic mammals are relatively hairless, others aren't. Some possess a significant layer of subcutaneous fat, others don't. Some are opportunistically bipedal, others aren't. There really is no single, fixed mammalian 'aqua-type' to which humans conform.

It is uncommon but not unheard of for terrestrial mammals other than humans to have the so-called *diving reflex* which appears specially evolved for aquatic life. However, the diving reflex can apparently be induced in the dog, which did not have an aquatic evolutionary phase. Accordingly, diving is not the only possible explanation for the diving reflex in humans. Hence, cataloguing human features that are analogous to those in certain aquatic or semi-aquatic mammals inevitably leads to some degree of selectivity. Physiological, morphological or behavioural features compatible with aquatic life are not enough evidence for evolution via water.

Could the diving reflex have arisen in humans because of something other than an aquatic phase in evolution? Perhaps its presence in humans has another explanation, i.e. arising from convergent evolution related more to respiratory control for speech than for diving. Like speech, that still makes the diving reflex fairly unique in humans and perhaps dependent on some aspect of a semi-aquatic phase in evolution but not necessarily because of diving. The same issues arise with other human features supposedly having evolved in an aquatic or, indeed, savannah environment. However, in the context of needing the shores for human brain evolution, other features of the human body that may or may not be so clearly dependent on shore-based evolution become more plausible.

Exploration by Sea

Notwithstanding the possibility that certain apparently shore-based adaptations like good underwater visual acuity or the diving reflex could have evolved in response to non-aquatic stimuli, their utility in the context of shore-based activities merits comment. The most obvious use *(exaptation)* that longterm exposure to inland or maritime shores would have for a bipedal hominid is to provide the opportunity to explore the coasts both on foot and by water. For instance, Australia was discovered at least 50,000 but perhaps as much as 80,000 years ago. To get to Australia required travel by water for a minimum of 50 miles (90 kilometers). Some form of watercraft was therefore needed that was seaworthy for a period of perhaps a week if not more.

Fifty to one hundred thousand years ago, the hominids that made it to Australia were probably *H. sapiens*, but trips of that distance would have been preceded by a great deal of short distance travel. Hundreds of thousands of years would have been needed to acquire the skills and confidence to explore to such distances. Longterm shore-based existence would have been an obvious prerequisite to such a process of familiarization with boat building and the hazards of ocean travel. More recently but equally challenging, the discovery of the Americas increasingly seems to have been via coastal routes from Northeast Asia.

At least 500,000 years ago, *H. erectus* became widely dispersed from Africa to the Middle East, China, and southeast Asia. Some of that exploration might have been by sea but much clearly must have occurred on foot. The acquisition of skills for ocean navigation by no means excludes other forms of exploration; on the contrary, they are complimentary because inland bodies of water on all the continents, especially large rivers, can present formidible obstacles to unassisted land-based exploration.

Hence the Shore-based Scenario is not only compatible with the major events in pre-human and early human evolution, it explains their circumstances more parsimoniously than other theories that exclude shore-based subsistence. The events of significance in human evolution start with the emergence of bipedalism and the evolution of fat babies and larger brains. They extend to the comparatively recent discovery of Australia and the Americas. Whether inland or on the seacoasts, these events were most plausibly a consequence of shore-based subsistence. Most importantly, the Shore-based Scenario offers a clear rationale for the most important hangover of human evolution – the ongoing developmental vulnerability that is peculiar to the brain.

Bibliography

Adam PAJ, Raiha N, Rahiala EL, Kekomaki EL. Oxidation of glucose and D-Beta-hydroxybuyrate by the early human fetal brain, *Acta Paediatr. Scand. 64*, 17-24, 1975.

Aeillo LC and Dean C. *Introduction to Human Evolutionary Anatomy.* Academic, New York, 1990.

Aiello LC, Wells JCK. Energetics and evolution of the genus *Homo. Ann Rev Anthropol 31*, 323-338, 2002.

Aeillo LC, Wheeler P. The expensive-tissue hypothesis. The brain and the digestive system in human and primate evolution. *Current Anthropol 36*, 199-221, 1995.

Allen JS, Damasio H, Grabowski TJ. Normal neuroanatomical variation in the human brain. An MRI-volumetric study. *Am J Phys Anthropol* 118, 341-358, 2002.

Andersen HT. Physiological adaptations in diving vertebrates. *Physiol Rev 46*, 212-243, 1966.

Armstrong E. Relative brain size and metabolism in mammals. *Science* 230, 1302-1304, 1983.

Arthur JR. Functional indicators of iodine and selenium status. *Proc Nutr Soc* 58, 507-512, 1999.

Balazs R, Brooksbank BWL, Davison AN, Eayrs JT, Wilson DA. The effect of neonatal thyroidectomy on myelination in the rat brain. *Brain Resarch 15*, 219-232, 1969.

Barker DJP. Maternal nutrition, fetal nutrition, and disease in later life. *Nutrition 13*, 801-813, 1997.

Barinaga M. What makes brain neurons run? *Science* 276, 196-198, 1997.

Barton RA, Harvey PH. Mosaic evolution in brain structure in mammals. *Nature* 405, 1055-1058, 2000.

Battaglia FC, Meschia G. *Fetal and Neonatal Physiology.* Cambridge University Press, Cambridge, UK, 1973.

Battaglia FC, Thureen PJ. Nutrition of the fetus and premature infant. *Nutrition 13*, 903-906, 1997.

Bazinet RP, McMillan EG, Seebaransingh R, Hayes AM, Cunnane SC. Whole body beta-oxidation of linoleate and alpha-linolenate in the pig varies markedly with weaning strategy and dietary alpha-linolenate. *J Lipid Res 44*, 314-319, 2003.

Benton D. Selenium intake, mood and other aspects of psychological functioning. *Nutr Neurosci* 5, 363-374, 2002.

Bliss TVP. Young receptors make smart mice. *Nature* 401, 25-27, 1999.

Blumenberg B. The evolution of the advanced human brain. *Current Anthropol* 24, 589-622, 1983.

Bogin B. Evolutionary hypotheses for human childhood. *Yearbook Physical Anthropol 40*, 63-89, 1997.

Bower JM, Parsons LM. Rethinking the 'lesser' brain. *Sci Amer* 289, 50-57, 2003.

Boyd Eaton S, Eaton SB, Sinclair AJ, Cordain L, Mann NJ. Dietary intake of long chain polyunsaturated fatty acids during the paleolithic. *World Rev Nutr Diet* 83, 12-23, 1998.

Brace CL, Montagu A. *Man's Evolution: An Introduction to Physical Anthropology*. MacMillan, London, 1965.

Bradshaw JL. *Encephalization and Growth of the Brain*. Ch. 8 (p. 145-156) in *Human Evolution and Neuropsychological Perspective*. Psychology Press, Hove, UK. 1997.

Brent GA, Moore DD, Larsen PR. Thyroid hormone regulation of gene expression. *Annu. Rev. Physiol 53*, 17-35, 1991.

Broadhurst CL, Cunnane SC, Crawford MA. Rift Valley lake fish and shellfish provided brain-specific nutrition for early *Homo*. *Brit J Nutr 79*, 3-21, 1998.

Broadhurst CL, Wang Y, Crawford MA, Cunnane SC, Parkington JE, Schmidt W. Brain-specific lipids from marine, lacustrine or terrestrial fodd resources; potential impact on early African *Homo sapiens*. *Comp. Biochem. Physiol. Part B 131*, 653-673. 2002.

Burton JH, Price TD. The ratio of barium to strontium as a paleodietary indicator of consumption of marine resources. *J Archeolog Sci 17*, 547-557, 1990.

Butte NF, Wong WW, Fiorotto M, O'Brien, Smith E, Garza C. Influence of early feeding mode on body composition of infants. *Biol Neonate 67*, 414-414, 1995.

Calvin WH. The emergence of intelligence. *Sci Amer*, 101-107, October 1994.

Chamberlain JG. The possible role of long chain omega 3 fatty acids in human brain phylogeny. *Persp Bio Med* 39, 436-445, 1996.

Changeux J-P and Chavaillon J. *Origins of the Human Brain*. Clarendon Press, Oxford, 1995.

Chisholm BS, Nelson DE, Schwarcz HP. Stable carbon isotope ratios as a measure of marine versus terrestrial protein in ancient diets. *Science 216*, 1131-1132, 1982.

Chou HH, Hayakawa T, Diaz S. Krings M, Indriati E, Leakey M, Pääbo S, Satta Y, Takahata N, Varki A. Inactivation of CMP-*N*- acetylneuraminic acid hydroxylase occurred prior to brain expansion during human evolution. *Proc Nat Acad Sci USA 99*, 11736-11741, 2002.

Conroy GC. *Reconstructing Human Origins: A Modern Synthesis*. WW Norton, New York, 1997.

Cooper DS. Subclinical thyroid disease: a clinician's perspective. *Ann Intern Med* 129, 135-138, 1998.

Cordain L, Watkins BA, Florant GL, Kelher M, Rogers L, Li Y. Fatty acid analysis of wild ruminant tissues. Evolutionary implications for reducing diet-related chronic disease. *Europ J Clin Nutr 56*, 1-11, 2002.

Cordain L, Watkins BA, Mann NJ. Fatty acid composition and energy density of foods available to African hominids. Evolutionary implications for human brain development. *World Rev. Nutr. Dietet. 90*, 144-161, 2001.

Clark DA, Mitra PP, Wang SS-H. Scalable architecture in mammalian brains. *Nature* 411, 189-193, 2001.

Crawford MA, Bloom M, Broadhurst CL, Schmidt W, Cunnane SC, Galli C. Evidence for the unique function of docosahexaenoic acid during the evolution of the modern hominid brain. *Lipids* 34, S39-S47, 1999.

Crawford MA, Costeloe K, Ghebremeskel K, Phylactos A, Skirvin L, Stacey F. Are deficits of arachidonic and docosahexaenoic acids responsible for the neural and vascular complications of preterm babies? *Amer J Clin Nutr S66*, 1032S-1041S, 1997.

Crawford, MA, Cunnane SC, Harbige LS. A new theory of evolution. Quantum Theory. P. 87-95 in *Essential Fatty Acids and Eicosanoids. Invited Papers from the Third International Congress.* Eds. Sinclair A, Gibson R. AOCS Press, Champaign, IL, 1992.

Crawford MA and Marsh D. *The Driving Force: Food in Evolution and the Future.* William Heinemann, London, 1989.

Cremer JE. Substrate utilization and brain development. *J Cerebral Blood Flow Metab 2*, 394-407, 1982.

Crockford SJ. *Animal Domestication and Heterochronic Speciation, The Role of Thyroid Hormone.* P. 122-153 in *Human Evolution through Developmental Change.* Eds Minught-Purvis N, McNamara KJ. The Johns Hopkins University Press, Baltimore, 2002.

Crockford S. Thyroid rhythm phenotypes and hominid evolution. A new paradigm implicates pulsatile hormone secretion in speciation and adaptation changes. *Comp Biochem Physiol* 135, 105-129, 2003.

Cronin JE, Boaz NT, Stringer CB, Rak Y. Tempo and mode in hominid evolution. *Nature* 292, 113-122, 1981.

Cross KW, Stratton D. Aural temperature of the newborn infant. *Lancet ii*, 1179-1180, 1974.

Cunnane SC. The aquatic ape theory reconsidered. *Med Hypoth 6*, 49-58, 1980.

Cunnane SC. New developments in alpha-linolenate metabolism with emphasis on the importance of beta-oxidation and carbon recycling. *World Rev Nutr Dietet 88*, 178-183, 2001.

Cunnane SC. Problems with essential fatty acids: Time for a new paradigm? *Progr Lipid Res 42*, 544-568, 2003.

Cunnane SC. Metabolism of polyunsaturated fatty acids and ketogenesis. An emerging connection. *Prostagl Leukotri Essent Fatty Acids 70*, 237-241, 2004.

Cunnane SC, Crawford MA. Survival of the fattest. Fat babies were the key to evolution of the large human brain. *Comp Biochem Physiol 136A*, 17-26, 2003.

Cunnane SC, Francescutti V, Brenna JT, Crawford MA. Breast-fed infants achieve a higher rate of brain and whole body docosahexaenoate accumulation than formula-fed infants not consuming dietary docosahexaenoate. *Lipids* 35, 105-111. 2000.

Cunnane SC, Harbige LS, Crawford MA. The importance of energy and nutrient supply in human brain evolution. *Nutr Health* 9, 219-235, 1993.

Cunnane SC, Menard CR, Likhodii SS, Brenna JT, Crawford MA. Carbon recycling into de novo lipogenesis is a major pathway in neonatal metabolism of linoleate and alpha-linolenate. *Prostagl Leukotr Essential Fatty Acids 60*, 387-392, 1999.

Cunnane SC, Ryan MA, Nadeau CR, Bazinet RP, Musa-Veloso, K, McCloy U. Why is carbon from some polyunsaturates extensively recycled into lipid synthesis. *Lipids 38*, 477-484, 2003.

Darwin C. *The Origin of Species*. Random House, Toronto, 1998.

Dawkins, R. *The Blind Watchmaker*. Penguin, London, 1986.

Deacon TW. What makes the human brain different? *Ann Rev Anthropol* 26, 337-357, 1997.

De Raveglia IF, Gomez CJ, Ghittoni N. Effects of thyroxine and growth hormone on the lipid composition of the cerebral cortex and the cerebellum of developing rats. *Neurobiology 3*, 176-184, 1973.

De Urquiza AM, Liu S, Sjoberg M, Zetterström RH, Griffiths W, Sjövall J, Perlmann T. Docosahexaenoic acid, a ligand for the retinoid X receptor in mouse brain. *Science 290*, 2140-2144, 2000.

Dobson JE. The iodine factor in health and evolution. *Geograph Rev 88*, 1-28, 1998.

Dussault JH and Ruel J. Thyroid hormones and brain development. *Ann Rev Physiol* 49, 321-334, 1987.

Eaton SB, Eaton SB, Konner MJ, Shostak M. An evolutionary pespective enhances understanding of human nutritional requirements. *J Nutr 126*, 1732-1740, 1996.

Eaton SB, Konner M. Paleolithic nutrition. A consideration of its nature and current implications. *New Engl J Med* 312, 283-289, 1985.

Edmond J. Ketone bodies as precursors of sterols and fatty acids in the developing rat. *J Biol Chem 249*, 72-80, 1974.

Edmonds CJ. Peripheral metabolism of thyroxine. *Journal of Endocrinology 114*, pp.337-39, 1987.

Ehrlich P, Feldman M. Genes and cultures: What creates our behavioral phenome? *Current Anthropol* 44, pp.87-107, 2003.

Ekstig B. Condensation of developmental stages and evolution. *BioScience* 44, 158-164, 1994.

Ellis D. Proboscis monkey and aquatic ape. *Sarawak Museum J 57*, 251-261, 1986.

Ellis DV. Wetlands or aquatic ape? Availability of food resources. *Nutr Health* 9, 205-217, 1993.

Elsner N, Schnitzler H-U. Eds. *Brain and Evolution. Proceedings of the 24th Gottingen Neurobiology Conference 1996.* Georg Thieme, Stuttgart, 1996.

Enard W, Khaitovich P, Klose J, Zöllner S, Heissig F, Giavalisco P, Nieselt-Struwe K, Muchmore E, Varki A, David R, Doxiadis GM, Bontrop RE, Pääbo S. Intra-and interspecific varitation in primate gene expression patterns. *Science* 296, 340-343, 2002.

Erecinska M, Silver IA. Iron and energy in mammalian brain. *Progr Neurobiol* 43, 37-71, 1994.

Evans RM. The steroid and thyroid hormone receptor superfamily. *Science* 240, 889-895, 1988.

Falk D and Gibson KR. *Evolutionary Anatomy of the Primate Cerebral Cortex.* Cambridge University Press, Cambridge, 2001.

Falk D. Hominid paleoneurology. *Ann Rev Anthropol* 16, 13-30, 1987.

Falk D. Hominid brain evolution. Looks can be deceiving. *Science* 280, 1714, 1998.

Farquharson J, Cockburn F, Patrick WA, Jamieson EC, Logan RW. Infant cerebral cortex phospholipid fatty acid composition and diet. *Lancet 340*, 810-813, 1992.

Farquharson J, Cockburn F, Patrick WA Jamieson EC, Logan RW. Effect of diet on infant subcutaneous tissue triglyceride fatty acids. *Arch Dis Child 69*, 589-593, 1993.

Finlay BL, Darlington RB. Linked regularities in the development and evolution of mammalian brains. *Science* 268, 1578-1584, 1995.

Francon J, Fellous A, Lennon AM, Nunez J. Is thyroxine a regulatory signal for neurotubule assembly during brain development? *Nature* 266, 188-190, 1977.

Gahtan V, Olson ET, Sumpio BE. Molecular biology. A brief overview. *J Vasc Surg* 35, 563-568, 2002.

Godfrey L, Jacobs KH. Gradual, autocatalytic and punctuational models of hominid brain evolution. A cautionary tale. *J Human Evol* 10, 255-272.

Gancedo B, Alonso-Gomez AL, de Pedro N, Delgado MJ, Alonso-Bedate M. Changes in Thyroid Hormone Concentrations and Total Contents through Onctogeny in Three Anuran Species: Evidence for Daily Cycles. *Gen Comp Endocrinol* 107, 240-250, 1997.

Geelhoed GW. Metabolic maladaptation: individual and social consequences of medical intervention in correcting endemic hypothyroidism. *Nutrition* 15, 11-12, 1999.

Gislen A, Dacke M, Kroger RHH, Abrahamsson M, Nilsson DE, Warrant EJ. Superior underwater vision in a human population of sea gypsies. *Current Biology 13*, 833-836, 2003.

Goldey ES, Kehn LS, Rehnberg GL, Crofton KM. Effects of developmental hypothyroidism on auditory and motor function in the rat. *Toxicol Appl Pharmacol* 135, 67-76, 1995.

Gould SJ. *Ever Since Darwin: Reflections in Natural History.* Burnett Books, London, 1978.

Gould SJ. *The Mismeasure of Man.* WW Norton, New York, 1981.

Gould SJ. *Wonderful Life.* Penguin, London 1989.

Gould SJ, Lewontin RC. Spandrels of San Marco and the Panglossian paradigm: a critique of the adaptationist programme. *Proc Royal Soc London B205*, 581-598, 1979.

Gunton JE. Iodine deficiency in ambulatory participants at a Sydney teaching hospital: is Australia truly iodine replete? *Med J Austral* 171, 467-470, 1999.

Hack M, Flannery DJ, Schluchter M, Cartar L, Borawski E, Klein N. Outcomes in young adulthood for very low birth weight infants. *New Engl J Med* 346, 149-157, 2002.

Hack M, Breslau N, Weissman B, Aram D, Klein N, Borawski E. Effect of very low birth weight and subnormal head size on cognitive abilities at school age. *New Engl J Med* 325, 231-237, 1991

Hack M, Taylor GH, Klein N, Eiben R, Schatschneider C, Mercuri-Minich N. School-age outcomes in children with birth weights under 750 g. *New Engl J Med* 331, 753-759, 1994.

Hak AE, Pols HAP, Visser TJ, Drexhage HA, Hofman A, Witteman JCM. Subclinical hypothyroidism is an independent risk factor for atherosclerosis and myocardial infarction in elderly women. The Rotterdam study *Ann Intern Med* 132, 270-278, 2000.

Hardy A. Was man more aquatic in the past? *New Scientist 7*, 642-645, 1960.

Hardy A. Was there a Homo Aquaticus? *Zenith 15*, 4-6, 1977.

Harrington TA, Thomas EL, Modi N, Frost G, Coutts GA, Bell JD. Fast and reproducible method for the direct quantitation of adipose tissue in newborn infants. *Lipids, 37*, 95-100, 2002.

Hille B. A potassium channel worthy of attention. *Science 273*, 1677, 1996.

Hirsch J, Farquhar JW, Ahrens, EH, Jr, Peterson ML, Stoffel W. Studies of adipose tissue in man. *Am J Clin Nutr 8*, 499-511,1960.

Hladik CM, Chivers DJ, Pasquet P. On diet and gut size in non-human primates and humans. Is there a relationship to brain size? *Current Anthropol* 40, 695-698, 1999.

Hofman M. Energy metabolism, brain size and longevity in mammals. *Quart Rev Biol* 58, 495-512, 1983.

Holliday M. Metabolic rate and organ size during growth from infancy to maturity and during late gestation and early infancy. *Pediatrics* 47, 169-172, 1971.

Hollowell JG, Staehling NW, Flanders WD, Hannon WH, Gunter EW, Spencer CA, Braverman LE. Serum TSH, T4, and thyroid antibodies in the United States population (1988-1994): National Health and Nutrition Examintion Survey (NHANES III). *J Clin Endocrinol Metab 87*, 489-499, 2002.

Holt AB, Cheek DB. Prenatal hypothyroidism and brain composition in a primate. *Nature* 243, 413-415, 1973.

Horrobin DF. Lipid metabolism, human evolution and schizophrenia. *Prostagl Leukotr Essent Fatty Acids* 60, 431-437, 1999.

Horrobin DF. *The Madness of Adam and Eve. How Shaped Humanity.* Bantam, New York, 2001.

Hotz CS, Fitzpatrick DW, Trick KD, L'Abbé MR. Dietary iodine and selenium interact to affect thyroid hormone metabolism of rats. *J Nutr* 127, 1214-1218, 1997.

Jeffery N, Spoor F. Brain size and the human cranial base: A prenatal perspective. *Amer J Phys Anthropol* 118, 324-340, 2002.

Jenkins, M. *Evolution.* Hodder Headline, London, 1999.

Jerison H. *Evolution of the Human Brain and Intelligence.* Academic Press, London, 1973.

Johanson D and Edey M. *Lucy: The Beginnings of Mankind.* Touchstone Books, New York. 1981.

Kahn P, Gibbons A. DNA from an extinct human. *Science* 277, 176-178, 1997.

Kappelman J. The evolution of body mass and relative brain size in fossil hominids. *J Human Evol* 30, 243-276, 1996.

Karten HJ. Evolutionary developmental biology meets the brain: The origin of mammalian cortex. *Proc Natl Acad Sci USA* 94, 2800-2804, 1997.

Kety SS. The general metabolism of the brain in vivo. P. 221-236 in *Metabolism of the Nervous System*, Ed. Richter D, Pergamon, London 1957.

Klein AH, Meltzer S, Kenny FM. Improved prognosis in congenital hypothyroidism treated before age three months. *J Pediatr* 81, 912-915, 1972.

Klein RG. Whither the Neanderthals? *Science* 299, 1525-1527, 2003.

Kondo K, Levy A, Lightman SL. Effects of maternal iodine deficiency and thyroidectomy on basal neuroendocrine function in rat pups. *J Endocrinol* 152, 423-430, 1997.

Krings M, Capelli C, Tschentscher, F, Geisert H, Meyer S, Von Haeseler A, Grossschmidt K, Possnert G, Paunovic M, Pääbo S. A view of Neandertal genetic diversity. *Nature Genetics* 26, 44-146, 2000.

Krings M, Geisert H, Schmits R, Krainitzki H, Pääbo S. DNA sequence of the mitochondrial hypervairable region II from the Neandertal type specimen. *Proc Natl Acad Sci* 96, 5581-5585, 1999.

Krings M, Stone A, Schmits RW, Krainitzki H, Stoneking M, Pääbo S. Neandertal DNA sequences and the origin of modern humans. *Cell* 90, 19-30, 1997.

Kuhn, TS. *The Structure of Scientific Revolutions.* 2nd Edition. University of Chicago Press, London, 1970.

La Lumiere LP. Evolution of human bipedalism. A hypothesis about where it happened. *Phil Trans Royal Soc Lond* 292, 103-107, 1981.

Langdon J. Umbrella hypotheses and parsimony in human evolution. A critique of the Aquatic Ape Theory. *J Human Evol* 33, 479-494, 1997.

Leakey R. *The Origin of Humankind.* Basic Books, New York, 1994.

Leakey RE and Lewin R. *People of the Lake: Mankind and Its Beginnings.* Anchor Press, New York, 1978.

Lee-Thorp J, Thackery JF, van der Merve N. The hunters and hunted revisited. *J Human Evol 39*, 565-576, 2000.

Leonard WR. Food for thought. Dietary change was a driving force in human evolution. *Sci Amer 287*, 106-115, 2002.

Leonard WR, Robertson ML. Evolutionary perspectives on human nutrition. The influence of brain and body size on diet and metabolism. *Am J Human Biol 6*, 77-88, 1994.

Leonard WR, Robertson ML. On diet, energy metabolism, and brain size in human evolution. *Current Anthropol 37*, 125-129, 1996.

Leonard WR, Robertson ML, Snodgrass JJ, Kuzawa CW. Metabolic correlates of human evolution. *Comp Biochem Physiol 136A*, 5-16, 2003.

Lewin R. How did humans evolve big brains? *Science 216*, 840-841, 1982.

Lewontin R. *The triple helix: Genes, organisms and the environment.* Cambridge, Harvard University Press, 2000.

Lewontin R. *It ain't necessarily so. The dream of the human genome and other illusions.* New York Review Books, New York, 2000.

Lindahl T. Facts and artifacts of ancient DNA. *Cell 90*, 1-3, 1997.

Lovejoy O. The origin of man. *Science 211*, 341-350, 1981.

Lozoff B, Brittenham GM. Behavioural aspects of iron deficiency. *Progr Hematol 14*, 23-53, 1986.

Lutz PL, Nilsson GE. *The Brain Without Oxygen*, 2nd Ed. Chapman and Hall, New York, 1997.

Lumsden CJ and Wilson EO. *Promethean Fire: Reflections on the Origin of Mind.* Harvard University Press, Cambridge, MA, 1983.

Mace GM, Harvey PH, Clutton-Brock TH. Brain size and ecology in small mammals. *J Zool 193*, 333-354, 1981.

MacKinnon, J. *The Ape Within Us.* Holt, Rinehart and Winston, New York, 1978.

Martin RD. Relative brain size and basal metabolic rate in terrestrial vertebrates. *Nature 293*, 57-60, 1981.

Martin RD. Scaling of the mammalian brain: the maternal energy hypothesis. *News Physiol Sci 11*, 149-156, 1996.

Martinez R, Toledano A. Histochemical study of the metabolism of ketone bodies in the nervous system. *Acta Histochem 38*, 218-226, 1970.

McCloy U, Ryan MA, Pencharz PB, Ross RJ, Cunnane SC. A comparison of the metabolism of eighteen carbon ^{13}C-unsaturated fatty acids in health women. *J Lipid Res 45*, 474-485, 2004.

McKinney ML. The juvenilized ape myth. Our 'overdeveloped' brain. *Bioscience 48*, 109-116, 1998.

Mink JW, Blumenschine RJ, Adams DB. Ratio of central nervous system to body metabolism in vertebrates: its constancy and functional basis. *Amer J Physiol* R203-R212, 1981.

Mitchell JH, Nicol F, Beckett GJ, Arthur JR. Selenium and iodine deficiencies: effects on brain and brown adipose tissue selenoenzyme activity and expression. *J Molec Endocrinol 155*, 255-263, 1997.

Mitchell JH, Nicol F, Beckett GJ, Arthur JR. Selenoprotein expression and brain development in pre-weanling selenium- and iodine-deficient rats. *J Molec Endocrinol 20*, 203-210, 1998.

Monastersky R. Children of the C4 world. Did a decline in carbon dioxide concentrations spur our evolution? *Science News 153*, 14-15, 1998.

Montagu A. *Human Heredity.* Second Edition. Signet Science Library Books, Toronto, 1963.

Montagu A. *Man: His First Million Years.* The New American Library, New York, 1960.

Morgan E. *The Descent of Woman.* Bantom Books, New York, 1972.

Morgan, E. *The Aquatic Ape: A Theory of Human Evolution.* Souvenir Press, London, 1982.

Morgan E. *The Scars of Evolution: What Our Bodies Tell Us About Human Origins.* Oxford University Press, New York, 1990.

Morgan E. *The Descent of the Child: Human Evolution from a New Perspective.* Oxford University Press, New York, 1995.

Morgan E. *The Aquatic Ape Hypothesis: The Most Credible Theory of Human Evolution.* Souvenir Press, London, 1997.

Morris D. *The Naked Ape.* Jonathon Cape, London, 1967.

Morris MS, Boston AG, Jacques PF, Selhub J, Rosenberg IH. Hyperhomocysteinemia and hypercholesterolemia associated with hypothyroidism in the third US National Health and Nutrition Examination Survey. *Atherosclerosis* 155, 195-200, 2001.

Nilsson GE. Brain and body oxygen requirements of *Gnathonemus Petersii*, a fish with an exceptionally large brain. *J Exper Biol* 199, 601-607, 1996.

Oldendorf WH, Cornford ME, Brown WJ. The large apparent work capability of the blood-brain barrier. A study of the mitochondrial content of capillary endothelial cells in brain and other tissue of the rat. *Ann Neurol* 1, 409-417, 1977.

Ovchinnikov I, Gotherstrom A, Romanove P, Kharitonov VM, Liden K, Goodwin W. Not just old but old and cold? *Nature* 410, 771-772, 2001.

Ovchinnikov IV, Gotherström A, Romanova GP, Kharitonov VM, Lidén K, Goodwin W. Molecular analysis of Neanderthal DNA from the northern Caucasus. *Nature, vol 404*. PP.490-493, 2000.

Pardridge WM. Blood-brain barrier transport of glucose, free fatty acids, and ketone bodies. P. 43-53 in *Fuel Homeostasis and the Nervous System*, ed. Vranic M *et al* Plenum, NY, 1991.

Parkington JE. The impact of the systematic exploitation or marine foods on human evolution. Colloquia of the Dual Congress of the International Association of the Study of Human Paleontology and the International Society of Human Biologists. Sun City, South Africa. June 28-July 4, 1998.

Passingham R. What's so special about man's brain? *New Scientist 27*, 510-511, 1975.

Passingham RE. Rates of brain development in mammals including man. *Brain Behav Evol* 26, 167-175, 1985.

Patel MS. Influence of neonatal hypothyroidism on the development of ketone-body-metabolizing enzymes in rat brain. *Biochem J* 184, 169-172, 1979.

Patel MS, Owen OE. Development and regulation of lipid synthesis from ketone bodies by rat brain. *J Neurochem 28*, 109-114, 1977.

Paulson OB, Newman EA. Does the release of potassium from astrocyte endfeet regulate cerebral blood flow? *Science* 237, 896-898, 1987.

Pennisi E. Did cooked tubers spur the evolution of big brains? *Science* 283, 2004-2005, 1999.

Pharaoah POD, Buttfield IH, Hetzel BS. Neurological damage to the fetus resulting from severe iodine deficiency during pregnancy. *Lancet* 13, 308-310, 1971.

Phinney SD. Ketogenic diets and physical performance. *Nutr Metab 1*, 2, 2004.

Picciano MF. Human milk: nutritional aspects of a dynamic food. *Biol Neonate* 72, 84-93. 1998.

Pinker S. *Revenge of the Nerds*. p. 149-210 in *How the Mind Works*. WW Norton, New York, 1997.

Pinker S. *The Blank Slate*. Penguin, New York, 2002.

Pleasure D, Lightman C, Eastman S, Lieb M, Abramsky O, Silberberg D. Acetoacetate and D-(-)-beta-hydroxybutyrate as precursors for sterol synthesis by calf oligodendrocytes in suspension culture. Extramitochondrial pathway for acetoacetate metabolism. *J Neurochem* 32, 1447-1450, 1979.

Pollitt E. Iron deficiency and cognitive function. *Ann Rev Nutr* 13, 521-537, 1993.

Pollitt E, Metallinos-Katsaras E. Iron deficiency and behaviour. Constructs, methods, and validity of the findings. *Nutr Brain* 8, 101-148, 1990.

Porterfield SP. Vulnerability of the developing brain to thyroid abnormalities: environmental insults to the thyroid system. *Environ Health Perspect* 102, suppl 2, 125-130, 1994.

Prechtl HFR. New perspectives in early human development. *Obstet Gynecol Reprod Biol 21*, 347-355, 1986.

Previc FH. Thyroid hormone in chimpanzees and humans. Implications for the origins of human intelligence. *Am J Phys Anthropol 118*, 402-403, 2002.

Raichle ME. The metabolic requirements of functional activity in the human brain: a positron emission tomography study. P. 1-4 in *Fuel Homeostasis and the Nervous System*, ed. Vranic M et al Plenum, New York, 1991.

Ramkrishnan U. Prevalence of micronutrient malnutrition worldwide. *Nutr Rev* 60, S45-S52, 2002.

Relethford JH. Genetics of modern human origins. *Annual Rev Anthropol* 27, 1-23, 1998.

Remer T, Neubert A, Manx F. Increased risk of iodine deficiency with vegetarian nutrition. *Brit J Nutr* 8, 45-49, 1999.

Rhys Evans PH. The paranasal sinuses and other enigmas. An aquatic evolutionary theory. *J Laryngol Otol 106*, 214-225, 1992.

Richards MP, Pettitt PB, Stiner MC, Trinkaus E. Stable isotope evidence for increasing dietary breadth in the European mid-Uppe Paleolithic. *Proc Natl Acad Sci USA, 98,* 6528-6532.

Robillard PY, Chaline J, Chaouat G, Hulsey TC. Preeclampsia/eclampsia and the evolution of the human brain. *Current Anthropol* 44, 130-134. 2003.

Roede M, Wind J, Patrick J, Reynolds V. *The Aquatic Ape: Fact or Fiction?* Souvenir Press, London, 1991.

Roncagliolo M, Garrido M, Walter T, Peirano P, Lozoff B. Evidence of altered central nervous system development in infants with iron deficiency anemia at six months. Delayed maturation of auditory brainstem responses. *Amer J Clin Nutr* 68, 683-690, 1998.

Routtenberg A, Cantallops I, Zaffuto S, Serrano P, Namgung U. Enhanced learning after overexpression of a brain growth protein. *Proc Nat Acad Sci USA* 97, 7657-7662, 2000.

Ruff CB, Trinkaus E, Holiday TW. Body mass and encephalization in Pleistocene *Homo. Nature* 387 173-176, 1997.

Sarda P, Lepage G, Roy CC, Chessex P. Storage of medium-chain triglyceride in adipose tissue of orally fed infants. *Am J Clin Nutr 45,* 399-405, 1987.

Schultz AH. *The Life of Primates.* Weidenfeld and Nicholson. 1969.

Schwarcz H. *Problems and limitations of absolute dating in regard to the appearance of modern humans in southwestern Europe.* P. 89-106 in *Conceptual Issues in Modern Human Origins Research,* Eds. Clark GA, Willermet CM. Walter de Gruyter, New York, 1997.

Schwarcz HP, Schoeninger MJ. Stable isotope analyses in human nutritional ecology. *Yearbook Phys Anthropol 34,* 283-321, 1991.

Sealy JC, van der Merwe N. Social, spatial and chronological patterning in marine food use a determined by ^{13}C measurements of Holocene human skeletons from the south-western Cape, South Africa. *World Archeol 20,* 87-102, 1988.

Shreeve, J. *The Neandertal Enigma: Solving the Mysteries of Modern Human Origins.* William Morrow, New York, 1995.

Smith RJ. Biology and size in human evolution. Statistical inference misapplied. *Current Anthropol 37,* 451-481, 1996.

Smith RJ. Estimation of body mass in paleontology. *J Human Evol* 43, 271-287, 2002.

Smith SV, Kroopnick P. Carbon-13 isotopic fractionation as a measure of aquatic metabolism. *Nature 294,* 252-253, 1981.

Sokoloff L: Measurement of local cerebral glucose utilization and its relation to local functional activity in the brain. P. 21-42 in *Fuel Homeostasis and the Nervous System,* ed. Vranic M et al Plenum, NY, 1991.

Sorcini M, Balestrazzi P, Grandolfo ME, Carta S, Giovanelli G. The national register of infants with congenital hypothyroidism detected by neonatal screening in Italy. *J Endocrinol Invest* 16, 573-577, 1993.

Sorensen MV, Leonard WR. Neandertal energetics and foraging efficiency. *J Human Evol 40*, 483-495, 2001.

Sponheimer M, Lee-Thorp JA. Isotopic evidence for the diet of an early hominid, Australopithecus africanus. *Science 283,* 368-370, 1999.

Sponheimer M, Lee-Thorp JA. Differential resource utilization by extant great apes and australopithecines. Towards solving the C4 conumdrum. *Comp Biochem Physiol 136A,* 27-34, 2003.

Staub JJ. Minimal thyroid failure: effects on lipid metabolism and peripheral target tissues. *Europ J Endocrinol 138,* 137-138, 1998.

Steudel-Numbers KL. The energetic cost of locomotion. Humans and primates compared to generalized endotherms. *J Human Evol* 44, 255-262, 2003.

Stewart KM. Early hominid utilisation of fish resources and implications for seasonality and behaviour. *J. Human Evol., 27*, 229-245, 1994.

Stewart K. A report on the fish remains from Beds I and II sites, Olduvai Gorge, Tanzania. *Darmst Beitrag Naturgesch 6,* 263-269, 1996.

Stringer C. Coasting out of Africa. *Nature 405,* 24-27, 2000.

Stringer C. Out of Ethiopia. *Nature* 423, 692-694, 2003.

Svennerholm L, Bostrom K. Changes in weight and composition of major membrane components of human brain during the span of adult life of Swedes. *Acta Neuropathol* 94, 345-352, 1997.

Swisher CC, Curtis GH and Lewin R. *Java Man: How Two Geologists' Discoveries Changed Our Understanding of the Evolutionary Path to Modern Humans.* Scribner, New York. 2000.

Tang Y-P, Shimizu E, Dube GR, Rampon C, Kerchner GA, Zhuo M, Liu G, Tsien JZ. Genetic enhancement of learning and memory in mice. *Nature* 401, 63-68, 1999.

Tattersall I. *Becoming Human: Evolution and Human Uniqueness.* Harcourt, Brace and Company. New York, 1998.

Tattersall I, Matternes JH. Once we were not alone. *Sci Amer,* 56-62, Jan 2000.

Tauber H. Carbon-13 evidene for dietary habits of prehistoric man in Denmark. *Nature 292,* 332-333, 1981.

Teaford MF, Ungar PS. Diet and the evolution of the earliest human ancestors. *Proc Natl Acad Sci USA 97*, 13506-13511, 2000.

Templeton AR. *Testing the Out of Africa Replacement Hypothesis with Mitochondrial DNA Data.* P. 329-360 in *Conceptual Issues in Modern Human Origins Research,* Eds. Clark GA, Willermet CM. Walter de Gruyter, New York, 1997.

Thurston JH, Hauhart RE, Dirgo JA, Levine SK, Jones EM, Ikossi-O'Connor MG, Pierce RW, Pollock PG. *Selected Blood and Brain Metabolites as an Index of Cerebral Energy Metabolism in Newborn and Developing Mice.* P. 271-281 in *Cerebral Metabolism and Neural Function,* Eds. Passonneau JV, Hawkins RA, Lust WD, Welsh FA. Williams and Wilkins, Baltimore, 1980.

Tobias PV. Some aspects of the multi-faceted dependence of early humanity on water. *Nutr. Health, 16,* 13-17, 2002.

Towers HM, Schulze KF, Ramakrishnan R, Kashyap S. Energy expended by low birth weight infants in the deposition of protein and fat. *Pediatr Res 41*, 584-589, 1997.

Ulijaszek SJ. *Energetics and human evolution.* P. 166-175 in *Human Energetics in Biological Anthropomtery.* Cambridge University Press, Cambridge, 1995.

Ulijaszek SJ. Comparative energetics of primate fetal growth. *Am J Hum Biol 14*, 603-608, 2002.

Van der Heyden JTM, Van Toor H, Wilson JHP, Hennemann G, Krenning EP. Effects of caloric deprivation on thyroid hormone tissue uptake and generation of low-T3 syndrome. *Amer J Physiol* E156-E163, 1986.

Van der Merve NJ, Thackeray JF, Lee-Thorp JA, Luyt J. The carbon isotope ecology and diet of *Australopithecus africanus* at Sterkfontein, South Africa. *J Human Evol 44*, 581-597, 2003.

Vanderpas JB, Contempré B, Duale NL, Goossens W, Bebe N, Thorpe R, Ntambue K, Dumont J, Thilly CH, Diplock AT. Iodine and selenium deficiency associated with cretinism in northern Zaire. *Amer J Clin Nutr 52*, 1087-1093, 1990.

Vanderpump MPJ, Tunbridge WMG, French JM, Appleton D, Bates D, Clark F, Grimly Evans J, Hasan DM, Rodgers H, Tunbridge F, Young ET. The incidence of thyroid disorders in the community. A twenty year follow-up of the Wickham Survey. *Clin Endocrinol 43*, 55-68, 1995.

Venturi S, Venturi M. Iodide, thyroid and stomach carcinogenesis. Evolutionary story of a primitive antioxidant? *Europ J Endocrinol 140*, 371-372, 1999.

Venturi S, Donati FM, Venturi M, Venturi M. Environmental iodine deficiency. A challenge to the evolution of terrestrial life. *Thyroid 10*, 727-729, 2000.

Verhaegen MJB. The aquatic ape theory and some common diseases. *Med Hypotheses 24*, 293-300, 1987.

Verhaegen M. Aquatic Ape Theory, speech origins and brain differenes with apes and monkeys. *Med Hypotheses 44*, 409-413, 1995.

Verhaegen M. Aquatic theory, the brain cortex and language origins. *ReVision 18*, 34-78, 1995.

Verhaegen M. Aquatic features in fossil hominids? P 75-114 in *The Aquatic Ape: Fact of Fiction?* Eds. Roede M, Wind J, Patrick JM, Reynolds V. Souvenir Press, London, 1991.

Verma M. Raghuvanshi RS. Dietary iodine intake and prevalence of iodine deficiency disorders in adults. *J Nutr Environ Med 11*, 175-180, 2001.

Vranic M et al. Editors, *Fuel Homeostasis and the Nervous System*, Plenum, NY, 1991.

Wade N. *The New York Times Book of Genetics.* The Lyon Press, Guilford, Connecticut, 1998.

Walrath D. Rethinking pelvic typologies and the human birth mechanism. *Current Anthropol 44*, 5-31, 2003.

Walravens P, Chase HP. Influence of thyroid on formation of myelin lipids. *J Neurochem 16*, 1477-1484, 1969.

Walter RC, Buffler RT, Bruggemann JH, Guillaume MMM, Selfe SM, Negassi B, Llibeski Y, Cheng H, Edwards RL, von Cosal R, Neraudeau D, Gagnon M. Early human occupation of the Red Sea coast of Eritrea during the last interglacial. *Nature* 405, 65-69, 2000.

Weetman AP. Hypothyroidism. Screening and subclinical disease. *Brit Med J* 314, 1175-1178, 1997.

Whiting BA and Barton RA. The evolution of the cortico-cerebellar complex in primates. Anatomical connections predict patterns of correlated evolution. *J Human Evol* 44, 3-10, 2003.

Widdowson EM, Changes in body proportion and composition during growth. P. 153-163 in *Scientific Foundations of Pediatrics*. Eds. Davies JA, Dobbing J. Heinemann, London, 1974.

Wiener R, Hirsch HJ, Spitzer JJ. Cerebral extraction of ketones and their penetration into CSF in the dog. *Am J Physiol* 220, 1542-1546, 1971.

Wildman DE, Uddin M, Liu G, Grossman LI, Goodman M. Implication of natural selection in shaping 99.4% nonsynonymous DNA identity between humans and chimpanzees: Enlarging genus *Homo*. *Proc Natl Acad Sci USA* 100, 7181-7188, 2003.

Willis D. *The Hominid Gang: Behind the Scenes in the Search for Human Origins.* Viking, New York, 1989.

Wood B. Hominid palaeobiology: Have studies of comparative development come of age? *Amer J Phys Anthropol* 99, 9-15. 1996.

Wynn M, Wynn A. Human reproduction and iodine deficiency: Is it a problem in the UK? *J Nutr Environ Med* 8, 53-64, 1998.

Zivin JA, Snarr JF. Glucose and D(-)-3-hydroxybutyrate uptake by isolated perfused rat brain. *J Appl Physiol* 32, 664-668, 1972.

Index